Impact of Climate-Change on Water Resources

Impact of Climate-Change on Water Resources

Editors

Christina Anagnostopoulou
Charalampos Skoulikaris

MDPI • Basel • Beijing • Wuhan • Barcelona • Belgrade • Manchester • Tokyo • Cluj • Tianjin

Editors
Christina Anagnostopoulou
Aristotle University of
Thessaloniki
Greece

Charalampos Skoulikaris
Aristotle University of
Thessaloniki
Greece

Editorial Office
MDPI
St. Alban-Anlage 66
4052 Basel, Switzerland

This is a reprint of articles from the Special Issue published online in the open access journal *Climate* (ISSN 2225-1154) (available at: https://www.mdpi.com/journal/climate/special_issues/water_resources).

For citation purposes, cite each article independently as indicated on the article page online and as indicated below:

LastName, A.A.; LastName, B.B.; LastName, C.C. Article Title. *Journal Name* **Year**, *Volume Number*, Page Range.

ISBN 978-3-0365-0110-9 (Hbk)
ISBN 978-3-0365-0111-6 (PDF)

© 2021 by the authors. Articles in this book are Open Access and distributed under the Creative Commons Attribution (CC BY) license, which allows users to download, copy and build upon published articles, as long as the author and publisher are properly credited, which ensures maximum dissemination and a wider impact of our publications.

The book as a whole is distributed by MDPI under the terms and conditions of the Creative Commons license CC BY-NC-ND.

Contents

About the Editors vii

Preface to "Impact of Climate-Change on Water Resources" ix

Charalampos Skoulikaris, Christina Anagnostopoulou and Georgia Lazoglou
Hydrological Modeling Response to Climate Model Spatial Analysis of a South Eastern Europe International Basin
Reprinted from: *Climate* **2020**, *8*, 1, doi:10.3390/cli8010001 1

Abdoulaye Oumarou Abdoulaye, Haishen Lu, Yonghua Zhu, Yousef Alhaj Hamoud and Mohamed Sheteiwy
The Global Trend of the Net Irrigation Water Requirement of Maize from 1960 to 2050
Reprinted from: *Climate* **2019**, *7*, 124, doi:10.3390/cli7100124 19

Yonas Dibike, Hyung-Il Eum, Paulin Coulibaly and Joshua Hartmann
Projected Changes in the Frequency of Peak Flows along the Athabasca River: Sensitivity of Results to Statistical Methods of Analysis
Reprinted from: *Climate* **2019**, *7*, 88, doi:10.3390/cli7070088 39

Matthijs van den Brink, Ymkje Huismans, Meinte Blaas and Gertjan Zwolsman
Climate Change Induced Salinization of Drinking Water Inlets along a Tidal Branch of the Rhine River: Impact Assessment and an Adaptive Strategy for Water Resources Management
Reprinted from: *Climate* **2019**, *7*, 49, doi:10.3390/cli7040049 57

Qingmin Meng
Climate Change and Extreme Weather Drive the Declines of Saline Lakes: A Showcase of the Great Salt Lake
Reprinted from: *Climate* **2019**, *7*, 19, doi:10.3390/cli7020019 71

Ngamindra Dahal, Uttam Babu Shrestha, Anita Tuitui and Hemant Raj Ojha
Temporal Changes in Precipitation and Temperature and their Implications on the Streamflow of Rosi River, Central Nepal
Reprinted from: *Climate* **2019**, *7*, 3, doi:10.3390/cli7010003 83

Asim Jahangir Khan, Manfred Koch and Karen Milena Chinchilla
Evaluation of Gridded Multi-Satellite Precipitation Estimation (TRMM-3B42-V7) Performance in the Upper Indus Basin (UIB)
Reprinted from: *Climate* **2018**, *6*, 76, doi:10.3390/cli6030076 99

Fotios Maris, Apostolos Vasileiou, Panagiotis Tsiamantas and Panagiotis Angelidis
Estimating the Future Function of the Nipsa Reservoir due to Climate Change and Debris Sediment Factors
Reprinted from: *Climate* **2019**, *7*, 76, doi:10.3390/cli7060076 117

About the Editors

Christina Anagnostopoulou is an Associate Professor of the Department of Meteorology and Climatology, School of Geology, Aristotle University of Thessaloniki (AUTh), Greece. She received her B.Sc. in Geology in 1996, her M.Sc. in Meteorology and Climatology in 1999, and her Ph.D. in Climatology in 2003. She worked as a research assistant from 2000 at AUTh. During the last decades, her research has been focused on climate extremes, synoptic and dynamic climatology, climatic changes and their impacts, and dynamical and statistical downscaling. She has more than one hundred publications (papers, book chapters, and contributions to international conferences) on the subject of climatology. She is an experienced researcher with involvement in more than 20 European and national researcher projects. She has also actively engaged in service activities within the scientific community, as a journal paper reviewer and conference organizer.

Charalampos Skoulikaris was born in Athens, Greece, in 1978. He obtained a diploma and a M.Sc. diploma in Electrical and Computer Engineering from the Democritus University of Thrace, Greece, in 2002 and 2004 respectively. He was awarded the title of Doctor by the Departments of Civil Engineering of the Aristotle University of Thessaloniki (AUTh), Greece, and the Département de Sciences de la Terre et de l'Environnement d'Ecole des Mines de Paris, ParisTech, France, in 2008. The same year, he was the elected the General Secretary of the UNESCO Chair and the International Network of Water-Environment Centres for the Balkans (INWEB) hosted by AUTh. His main research and teaching interests are related to the sustainable and integrated management of water resources at the river basin scale, hydrology and hydrological modelling, simulation of hydroelectric projects, assessment of climate change and integration to water resources, climate change adaptation, and the estimation of environmental costs and socio-economic impacts of large-scale projects in river basins. He has authored approximately 75 scientific papers, book chapters, and contributions to international conferences, and he has worked as a primary researcher on more than 25 European and national research projects.

Preface to "Impact of Climate-Change on Water Resources"

The management and allocation of water resources is not a forthright procedure. Water users and stakeholders claim these resources for coverage of their own demands, such as domestic and industrial supply, irrigated agriculture, hydropower production, and ecosystem preservation. In cases of transboundary water resources, differentiations in national strategies, development priorities, and economic status among countries that share these resources induce more complexity in the management of water. Moreover, this complication may be further affected due to the demographic and climatic change drivers that increase the stress on water resources.

According to the IPCC, climate change is expected to have dramatic impacts on water resources and their management. Climate model simulations for the 21st century are consistent in projecting temperature increases resulting in water temperature augmentation, sea level rise, and thus changes in coastal regions. Higher water temperatures and changes in extremes, including floods and droughts, are projected to affect water quality and exacerbate many forms of water pollution. To better understand the mechanisms of climate variability and climate change on water resources, it is crucial to implement multidisciplinary studies that involve climatology and hydrology.

The articles presented in this book highlight the impact of climate change on water resources in different regions of the world and at different scales; from the catchment to the region and to the global scale. The reader will be informed on the way that changes in temperature and rainfall impact a wide range of hydrological processes including drought, streamflow, and irrigation water availability or even salinization.

The aim of the eight papers selected in this volume is to present a comprehensive picture of climate change impacts on water resources in case studies that are governed by different climatic characteristics. Furthermore, our intention was also to include articles that emphasize the diversity and complexity of hydrologic processes together with the climatic complexity at different environments, as well as to shed light on the coupling of climate projections with water-related simulation models. Most of the included research articles are supported by national and international funding, an issue that indicates the importance that is given to the theme of climate change and in particular to its impact on water resources.

The authors are grateful to the editors, reviewers, and production team. We hope that this Special Issue will foster additional advancements on the impacts of climate change on water resources and that the presented methodological approaches and outputs will be used for further research initiatives and applications.

Christina Anagnostopoulou, Charalampos Skoulikaris
Editors

Article

Hydrological Modeling Response to Climate Model Spatial Analysis of a South Eastern Europe International Basin

Charalampos Skoulikaris [1],*, Christina Anagnostopoulou [2] and Georgia Lazoglou [2]

[1] Department of Civil Engineering, Aristotle University of Thessaloniki, 54124 Thessaloniki, Greece
[2] Department of Meteorology Climatology, School of Geology, Aristotle University of Thessaloniki, 54124 Thessaloniki, Greece; chanag@geo.auth.gr (C.A.); glazoglou@geo.auth.gr (G.L.)
* Correspondence: hskoulik@civil.auth.gr; Tel.: +30-2310-995666

Received: 24 October 2019; Accepted: 18 December 2019; Published: 19 December 2019

Abstract: One of the most common questions in hydrological modeling addresses the issue of input data resolution. Is the spatial analysis of the meteorological/climatological data adequate to ensure the description of simulated phenomena, e.g., the discharges in rainfall–runoff models at the river basin scale, to a sufficient degree? The aim of the proposed research was to answer this specific question by investigating the response of a spatially distributed hydrological model to climatic inputs of various spatial resolution. In particular, ERA-Interim gridded precipitation and temperature datasets of low, medium, and high resolution, i.e., $0.50° \times 0.50°$, $0.25° \times 0.25°$, and $0.125° \times 0.125°$, respectively, were used to feed a distributed hydrological model that was applied to a transboundary river basin in the Balkan Peninsula, while all the other model's parameters were maintained the same at each simulation run. The outputs demonstrate that, for the extent of the specific basin study, the simulated discharges were adequately correlated with the observed ones, with the marginally best results presented in the case of precipitation and temperature of $0.25° \times 0.25°$ spatial analysis. The results of the research indicate that the selection of ERA-Interim data can indeed improve or facilitate the researcher's outputs when dealing with regional hydrologic simulations.

Keywords: hydrologic modeling; reanalysis gridded datasets; ERA-Interim; Balkan Peninsula

1. Introduction

The accuracy of hydrologic models is limited by many factors [1,2]. Data availability, both in terms of quantity, i.e., large data series and spatial coverage of the case study basin, and quality, i.e., reliable and unbiased datasets, plays a significant role in the modeling procedure and is one of the factors that are bound to affect the produced results. O'Riordan [3] demonstrated that the lack of historic data or even the comprehensiveness of monitoring could lead to distorted findings. In hydrological simulations and forecasting, precipitation is one of the most important inputs, with the precipitation's gauge network density and the gauges' spatial distribution having direct impacts on the modeling results [4]. Xu et al. [5] showed that the error of the simulated runoff was gradually narrowed to a specific threshold number of gauges, beyond which the model's performance did not demonstrate considerable improvements. Similarly, Anctil et al. [6] demonstrated that a model's performance was reduced when spatial rainfall was derived from a network where the number of network gauges was lower than a specific threshold. Woods et al. [7] found that the size of the representative elementary area [8] was influenced more by the catchment topography than by the spatial resolution of the rainfall data. Moreover, the density of the stations/grids for a hydrologic network that was defined by the World Meteorological Organization [9] indicated that the density was strongly dependent on the physiographic characteristics of the regions. The current availability of gridded precipitation data series [10–12] offers spatial coverage at various resolutions on a worldwide scale.

The accuracy of gridded data sources was thoroughly examined in the literature [13–16]. In northeast China, for example, a comparative study regarding gridded datasets, such as those of the Global Precipitation Climatology Center (GPCC), the Climate Research Unit (CRU), and the University of Delaware (UDEL), with station-based precipitation data, demonstrated that gridded databases overestimated the annual precipitation [17]. At the European scale, the comparison of existing precipitation datasets with E-OBS gridded data revealed that the differences were relatively large, and usually biased toward lower values in E-OBS [18]. Nevertheless, the selection of the proper grid resolution is of particular significance, since low grid resolution could lead to uncertainty of the predictions, while, in the case of overestimation of resolution, the workload is increasingly demanding [19]. Gridded climatic variables were also exploited for the assessment of other atmospheric processes, such as the daily global solar radiation [20].

Meteorological reanalysis data are among the most used gridded datasets, with those most widely used presented by Fuka et al. [21]. Although both gridded data and reanalysis data initially come from terrestrial and airborne observation networks, the reanalysis data go a step forward, i.e., they are assimilated into a numerical weather prediction model to produce a spatially and temporally coherent synthesis of meteorological variables covering the last few decades [22]. Reanalysis provides a multivariate, spatially complete, and coherent record of the global atmospheric circulation [23], and its usefulness was proven both in areas where there is a plethora of data and in areas where weather stations are limited or even do not exist [24]. Many studies compared reanalysis products to observation data and, in general, they concluded that reanalysis products are comparable to station measurements [23,25,26]. ERA-Interim is one of the latest global reanalysis products developed by the European Center for Medium-Range Weather Forecasts (ECMWF) [23]. ERA-Interim is highly used over regions with sparse observations, such as high mountainous regions or complex terrains [27,28]. The accuracy of these datasets was evaluated in many places of the word. For example, the Hu et al. [29] investigation regarding the reliability of the precipitation variable took place in central Asia. ERA-Interim was also used to assess, in terms of consistency, the temperature and precipitation extremes [30].

One of the principal hydro-climate applications of gridded datasets is to incorporate them into spatially distributed hydrologic models as climate forcings [31]. The literature presents some recent studies that investigated the impact of the spatial resolution of reanalysis products on hydrologic modeling [32,33]. In a large mountain watershed in Canada, for example, Woo and Thorne [34] exploited ERA-40, NCEP–NCAR, and NARR reanalysis products to simulate the contribution of snowmelt to the river regime. However, coupling of reanalysis data with hydrologic models was less explored and, in general, studies on this topic focused on a limited number of basins [22]. Moreover, the literature review showed that the coupling of reanalysis data with hydrologic models is mainly based on single spatial resolutions [35,36]. Fuka et al. [21], for example, in their hydrologic simulation, used specific reanalysis products covering the globe at hourly time steps since 1979 at a 38-km resolution. Limited researches evaluated a broader range of spatial analysis, such as Essou et al. [22], where they used three different reanalysis datasets of spatial resolution varying between 30 km and 10 km to trigger the hydrologic simulation procedure.

Based on the aforementioned review, this specific research aims at investigating the runoff response of a watershed in reanalysis climatic data of varying resolution. In particular, this study seeks to assess the sensitivity of a hydrologic model, namely MODSUR, by comparing the low-resolution ERA-Interim datasets ($0.50° \times 0.50°$) with the medium-resolution ERA-Interim datasets ($0.25° \times 0.25°$) and the high-resolution ERA-Interim datasets ($0.125° \times 0.125°$). The case study area is a transboundary river basin in southeastern Europe (SEE), where the climatic conditions in the upstream part of the basin are different to those of the downstream part due to its proximity to the sea. The performed analysis on the three different datasets demonstrated the degree of correlation among the relevant climatic variables, as well as the correlation of the simulated discharges with the observed discharges of the river. This research is considered of significant importance because ERA-Interim reanalysis data are routinely used for (i) case areas where the lack of data is dominant, and (ii) the bias correction of climatic variables, when climate change is inserted into the research. Hence, the outputs could shed light on questions relative to the required resolution of climatic data when used in hydrologic studies.

2. Materials and Methods

2.1. Case Study Area

The transboundary basin of Mesta/Nestos River, which is shared between Bulgaria and Greece, was the area of interest in this research. The specific basin is one of the 18 international rivers and lake basins in southeastern Europe and one of the five transboundary river basins of Greece. Moreover, this basin forms part of the Hydrology for the Environment, Life and Policy (HELP) demonstration basins of UNESCO's Intergovernmental Hydrological Programme (IHP) [37]. The morphology of the catchment is mountainous with the exception of the delta region (Figure 1). Due to the basin's (i) orientation from north to south and (ii) complex topography, since it spreads among the highest mountains of SEE and the sea level, the climate can change from typically coastal Mediterranean to practically alpine [38].

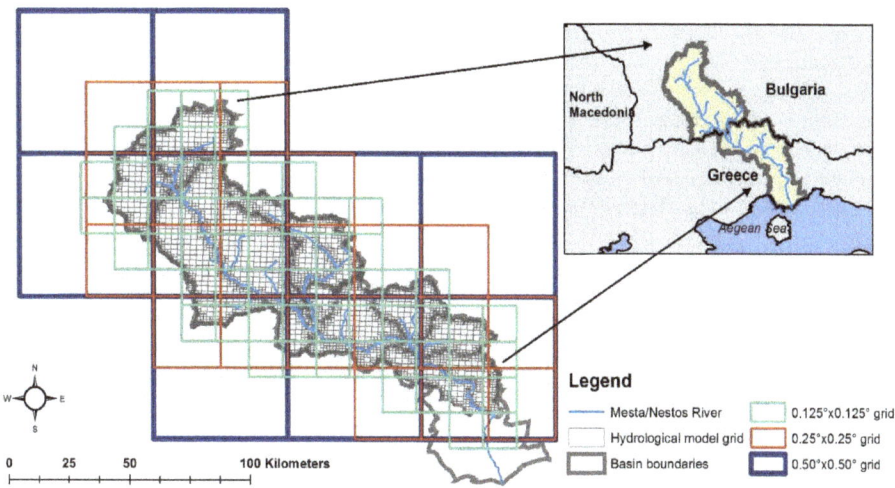

Figure 1. Case study basin with overlaying grids of the hydrologic model (gray rectangles) and of the 0.50° × 0.50° (blue cells), 0.25° × 0.25° (red cells), and 0.125° × 0.125° (green cells) ERA-Interim meshes.

The headwaters are located in southwestern Bulgaria, while the river outlets are located in the north Aegean Sea, in Greece. The total length of the river's main course and the basin's extent are 255.0 km and 6,218.0 km², respectively, figures that are almost equally shared by the two countries (Figure 1). The average inflows into Greece are estimated at 0.14×10^9 m³, i.e., water volumes that are significantly lower (approximately 50%) than the 0.278×10^9 m³ that are referred to in the Water Convention [39]. The observed decrease in water inflows from the upstream country to the downstream country are mainly attributed to the climatic variations. Bulgaria, in particular, argues [39] that, over the last 20 years, precipitation presented a decrease of 30%, thus leading to a subsequent decrease in water discharge.

2.2. Reanalyis Data and Derived Datasets

The analyzed data consisted of temperature and precipitation time series obtained from ERA-Interim, produced by the ECMWF (European Center for Medium-Range Weather Forecasts). Detailed information about the ERA-Interim reanalysis can be found in Dee et al. [23]. The precipitation data used in this study were projected on a grid of 0.5° × 0.5° (ERAI_50), 0.25° × 0.25° (ERAI_25), and 0.125° × 0.125° (ERAI_12.5) from the original Gaussian reduced grid (T255 reduced Gaussian grid of about 0.7° × 0.7°) [40], and they are provided at a daily time step. The length of the time series was 35 years extending from 1981 to 2015, with the specific length considered sufficient to carry out statistical analysis of the climatic data. ERA-Interim is an improved version compared to previous reanalysis products of ECMWF, based on the use of additional observations, the updated data assimilation system, and the increased resolution [23].

2.3. River Basin Simulation

The simulation of the basin's discharges was accomplished with the MODSUR (modélisation de transfers de surface) spatial distributed hydrologic model. MODSUR is the surface modeling component of the coupled surface and groundwater MODCOU (modélisation couplée) simulation model, which was developed by the Ecole Nationale Supérieure des Mines de Paris [41]. The model operation is based on a densely spaced grid, and it uses a progressive quadtree structure with varying cell sizes. This means that the surface domain is divided into grid cells of size a, 2a, 4a, and 8a, with the higher resolution attributed to the grid cells representing the river. For the selected case study basin, the utilized grid was composed of 9212 cells. The ensemble of connected cells builds the runoff network, which gathers the flow down to the catchment outlet. The water budget in the model was computed in each grid cell using a system of four reservoirs [41,42], as a function of precipitation (P), evapotranspiration (ETR), and water level of the reservoir (R), i.e., the initial stocked water in the soil. The system of reservoirs is responsible for the repartition of rainfall water into runoff, infiltration, evapotranspiration, and soil water storage (Figure 2).

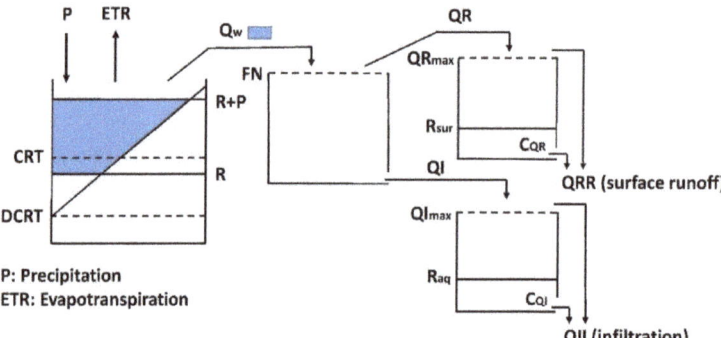

Figure 2. Representation of the MODSUR (modélisation de transfers de surface) hydrologic model operation mode.

The excess water transferred from the first to the second reservoir is defined as follows:

$$Q_w = max(R + P - R_{max}, 0) + \frac{dR(2RBA + dR)}{4(CRT - DCRT)}, \qquad (1)$$

where *DCRT* is the minimum stocked water (mm) in the soil below which no water quantity is available, and *CRT* the average stocked water quantity (mm), while R_{max}, *RBA*, and *dR* are expressed as follows:

$$\begin{aligned} R_{max} &= 2(CRT\text{--}DCRT) + DCRT; \\ RBA &= max(DCRT, R)\text{--}DCRT; \\ dR &= max(0, RHA - RBA); \\ RHA &= min(R + P, R_{max}) - DCRT. \end{aligned} \quad (2)$$

The partitioning of the water to runoff (*QR*) and infiltration (*QI*) is conducted in the second reservoir. It is controlled by the parameter *FN* that corresponds to the maximum value of infiltration over a time step (mm/day), and it is expressed as follows:

$$\begin{aligned} &\text{If } Q_w < FN, \text{ then } QI = Q_w \text{ and } QR = 0, \\ &\text{If } Q_w > FN, \text{ then } QI = FN \text{ and } QR = Q_w - FN. \end{aligned} \quad (3)$$

The third and fourth reservoirs are responsible for the calculation of the final infiltration (*QII*) and the surface runoff (*QRR*). The latter is further analyzed as pure runoff (if overflowing) and delayed runoff. The final calculation of the *QRR* is given as follows:

$$\begin{aligned} R_{sur} &= R_{sur} + QR, \\ &\text{If } R_{sur} < QR_{max}, \text{ then } QRR = C_{QR} \times R_{sur}, \\ &\text{If } R_{sur} \geq QR_{max}, \text{ then } QRR = C_{QR} \times R_{sur} + (QR - R_{sur}), \end{aligned} \quad (4)$$

where R_{sur} is the level of the surface runoff reservoir (mm), QR_{max} the surface runoff reservoir's overflow level (mm), and C_{QR} is the depletion ratio of the surface runoff reservoir (mm). The operation of the infiltration reservoir is similar to the aforementioned reservoir. For the specific case study, no interactions between the surface domain and the water table were introduced due to lack of data. Thus, the infiltration reservoir was not inserted in the simulation process.

For the Mesta/Nestos basin, the reservoir parameters were based on the catchment characteristics with digitized maps of geology and land uses to be overlaid on the hydrologic grid. For each cell of the grid, the dominant characteristics, e.g., geological formations and land-use types, were selected to define the relevant infiltration and evapotranspiration coefficients per cell. The MODSUR model was implemented in basins of varying scales, e.g., the Maritza basin in Bulgaria [43], as well as the selected case study [37].

Because of the increased altimetry of the basin, as the highest peak of the Balkan Peninsula (namely, peak Musala of 2925 m above sea level) is located within the basin, the snow coverage and melting processes play a significant role both in the river's spring increased discharge and in the continuity of the discharge during the summer. The snow component NEIGE, which is a compatible add-on of the MODSUR model, was used to simulate the snow cover regime on the principle of "degree days" [44,45], using an approach which distinguishes snow melting processes between forested and non-forested areas. The degree-day method is a temperature index method that equates the total daily melt to a coefficient times the temperature difference between the mean daily temperature and a base temperature (generally 0 °C).

$$M = C_M(T_a - T_b), \quad (5)$$

where *M* is the snowmelt expressed in mm/day, C_M is the degree-day coefficient (mm/degree-day °C), and T_a and T_b are the mean daily air and base temperatures (°C), respectively. The coefficient C_M depends on the season and the location, and it varies between 1.6 and 6.0 mm/degree-day °C [45].

In the NEIGE model [46], the snowmelt process is conditioned by the following equation:

$$t_{sto}(j) = ((1-cof) \times t_{sto}(j-1) + cof \times t_{mean}(j)) > t_s, \quad (6)$$

where t_{sto} is the temperature of the stocked snow (°C), cof is a coefficient of warming of the stocked snow, t_{mean} is the average daily air temperature (°C), and t_s is the threshold temperature for the snow melting (°C).

If this first condition is verified, then a portion of the stocked snow layer could be melted. However, the stocked snow layer has a specific storage capacity of liquid water. In order for the water to be outflowed, the volume of the stocked water should exceed that storage capacity, with the latter expressed as follows:

$$S_{nt}(j) = S_{nt}(j-1) + t_{tf} \times (t_{sto}(j) - S_{ts}) > S_n(j), \quad (7)$$

where S_{nt} is the water (in liquid form) accumulated in the stocked snow layer (mm), t_{tf} is the percentage of transformation (mm/°C), S_{ts} is the threshold temperature for the transformation (°C), and S_n is the stocked snow (mm). When the previous conditions are coupled, then the quantity, $F_n(j)$, of the melted snow in mm is given as follows:

$$F_n(j) = min\left[S_n(j), t_{tf} \times (t_{mean}(j) - t_s)\right]. \quad (8)$$

For the case study basin, the parameters of the NEIGE model, such as the threshold temperatures in °C for the snow melting in the forested and non-forested areas, were retrieved by Etchevers and Martin [47]. The variables proposed by the aforementioned authors were used by the French Meteorological Organization in their snow melt model for a Mediterranean basin, i.e., a basin which has similar characteristics to the current case study.

As aforementioned, the MODSUR model was already successfully applied to the study area, with the calibration and validation period from 1987 to 1993 (R^2 = 0.64) and from 1994 to 1995 (R^2 = 0.68), respectively [37]. For the model calibration, the precipitation data came from a network of 18 stations covering both parts of the basin, while the measured discharges were derived from two gauge stations. Although the rainfall datasets were available at a daily time step, the time step of the historical discharges was at the monthly level. For that purpose, the daily simulated outputs of the hydrologic model were averaged at monthly mean values in order to perform the model validation. In the present research, the parameterization of the model was the same as the aforementioned one [37], and the modifications were related to the forcing variables. Daily precipitation and temperature came from the three different climatic datasets, one per climate model, and they covered a period of 15 years, i.e., from 1 January 1981 to 31 December 1995. The potential evapotranspiration (PET) was calculated based on the Thornthwaite method utilizing the temperature time series of each dataset. The climatic gridded variables were nested in the hydrologic model grid with the use of Geographic Information Systems' (GIS) spatial analyst tools. Finally, in order to perform a comparison with the historical monthly observations, the simulated daily discharges for each one of the ERA-Interim datasets were aggregated at monthly level. It should also be stated that the simulated flows using different ERA-Interim resolution datasets were conducted on a 15-year period (1981–1995) that included the nine-year calibration validation period (1987–1995) of the hydrologic model.

The evaluation of the model's results was conducted with the coefficient of determination (R^2), which describes the degree of collinearity between simulated and measured data, as well as the percent bias (PBIAS). The latter assesses the average tendency of the simulated dataset to be either smaller or larger than the observed corresponding items [48].

3. Results

In the following two sections, the analysis of both annual and seasonal temperature and precipitation of 35 years, i.e., from 1981 to 2015, is presented. The spatial variability of the temperature and precipitation variables was explored based on the rotated principal component analysis (RPCA). By using the RPCA, newly projected data were acquired by transforming (rotating) all the original variables with the principal components [49]. Two rotated components were explored for both temperature and precipitation annual data representing climatic variability over the northern (N) and southern (S) parts of the river basin. The last section is devoted to the comparable analysis of the basin discharges that were produced by the coupling of the three different reanalysis datasets with the hydrologic model.

3.1. Climate Analysis of Temperature Data

The annual temperature line plots for the northern (solid lines) and southern (dashed lines) Mesta/Nestos catchment are demonstrated in Figure 3. The different colors attribute the results of ERAI_50 (gray-colored curves), ERAI_25 (yellow-colored curves), and ERAI_12.5 (blue-colored curves) spatial analysis. As shown in the figure, there was a difference of almost 2.0–3.0 °C between the temperature of the northern and the southern parts of the basin, since the northern part is dominated by high mountains while the southern part gets gradually flatter upon approaching to the sea. The temperature difference was detected for all the evaluated data resolutions. Additionally, the data with the highest resolution (0.125° × 0.125°) presented the lowest temperature in both areas, while the data of 0.50° × 0.50° resolution were associated with the highest temperatures. The temperature for ERAI_25 varied between the temperatures of high- and low-resolution datasets.

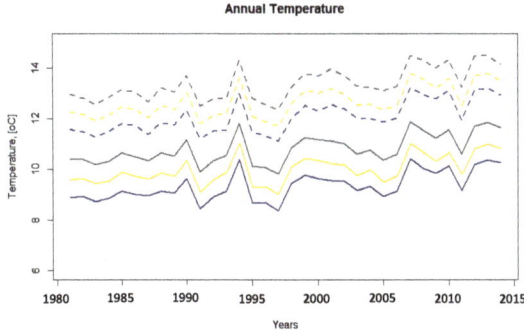

Figure 3. Line plot of annual temperature data for the northern (solid lines) and southern (dashed lines) parts of the Mesta/Nestos basin. Temperatures at 0.50° × 0.50°, 0.25° × 0.25°, and 0.125° × 0.125° resolution are represented in gray, yellow, and blue, respectively.

The seasonal diagram, as depicted in Figure 4, shows that the southern part of the basin had the highest temperatures during the whole year (all seasons). Additionally, the temperatures (blue colored lines) derived from the higher-resolution spatial analysis datasets were the lowest in both areas and during all seasons, while those derived from the coarser-resolution datasets were the highest (gray-colored lines). The most evident temperature bias between the two regions was detected during summer, with the southern flat region warmer than the northern mountainous region by about 2.0–3.0 °C. According to the seasonal graphs, the second warmest season was spring (MAM: March–April–May). The corresponding boxplots describe the difference between the northern and the southern parts of the Mesta/Nestos basin for the three spatial resolutions (N50–S50: gray; N25–S25: yellow; N12.5–S12.5: blue).

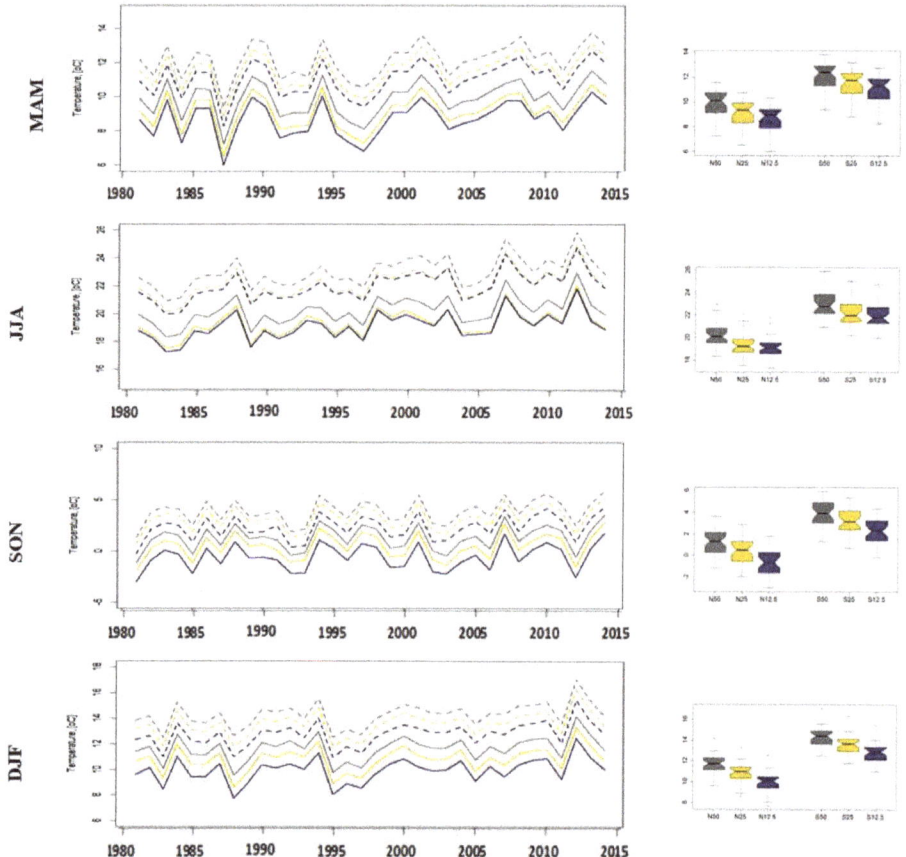

Figure 4. Line plots and boxplots of seasonal temperature data for the northern (solid line) and southern (dash line) parts of the basin. The temperatures at 0.50° × 0.50°, 0.25° × 0.25°, and 0.125° × 0.125° resolution are represented in gray, yellow, and blue, respectively.

3.2. Climate Analysis of Precipitation Data

The results for the precipitation parameter show that the southern part of the basin, which is the warmest, was also the driest, since its precipitation was almost 100 mm lower than that of the upstream part. The most important output for the precipitation variable was that all the datasets with different spatial resolution presented almost equal results. The statistics of the annual results can be seen in Table 1, where the values of the mean, maximum, interquartile range (IQR), standard deviation (SD), skewness, and kurtosis are presented for the two parts of the basin and for all resolutions. According to the analysis, the mean, maximum, and IQR values were higher in the northern part of the basin, e.g., the mean precipitation of N50 and S50 (N50 and S50 stands for the northern and southern parts of basin with data at 0.50° × 0.50°, respectively) was equal to 2.3 mm and 1.9 mm, respectively. The skewness and kurtosis were lower in the northern than in the southern part, while the standard deviation was almost equal in both parts of the basin. The results between the different spatial analyses in the same area presented no differences. As it can be seen in Table 1, in both areas, the climatic characteristics of precipitation were very close, meaning that the lowest-resolution spatial analysis could provide the same accuracy as the highest one.

Table 1. Statistics of the precipitation data at 50.0 × 50.0, 25.0 × 25.0, and 12.5 × 12.5 km resolution for the northern and southern parts of the Mesta/Nestos basin. IQR—interquartile range.

	N50	N25	N12.5	S50	S25	S12.5
Mean	2.3	2.4	2.4	1.9	1.9	1.9
Maximum	60.9	60.1	59.3	72.5	69.8	69.6
IQR	2.8	2.9	2.9	1.5	1.7	1.7
SD	4.4	4.5	4.5	4.5	4.4	4.4
Skewness	3.6	3.5	3.4	4.6	4.5	4.5
Kurtosis	18.7	17.5	17.0	31.1	30.2	30.4

The corresponding seasonal results for the two basin regions and for the three spatial resolutions are demonstrated in Figure 5. In both parts of the basin, the three spatial analyses had almost equal results during the four seasons and during the 35 years of data availability. The results were equal in most years except for some very specific cases. For example, in 1982, in both areas, the analysis of the ERA_12.5 data presented slightly higher precipitation for the summer season (JJA: June–July–August), while, in the year 2000, the data derived from the lowest-resolution dataset presented a slightly lower precipitation during summer (north part of the basin) and slightly more rainfall during autumn (southern part of the basin). It should be mentioned that, after 2003, the wet seasons were more obvious in both parts of the basin, while the previous years were characterized by drought conditions, particularly the years of 1984, 1986, and 1992 for both parts of the basin, while the years 2000 and 2001 were more intense in the southern part of the basin.

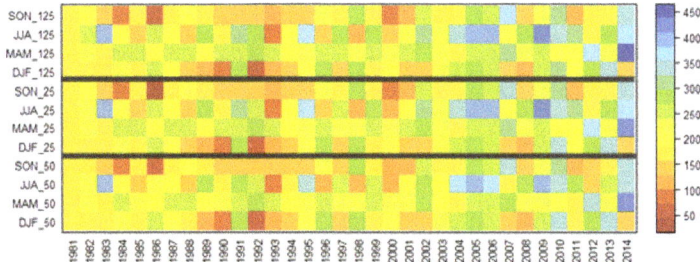

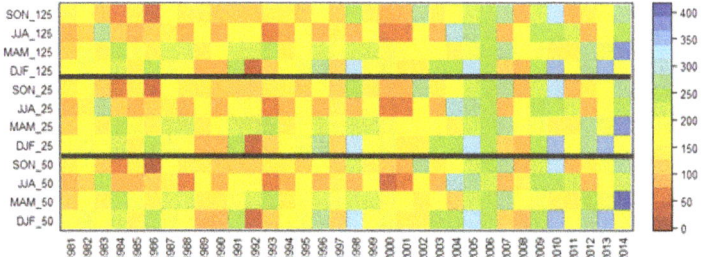

Figure 5. Seasonal precipitation data for north and south Mesta/Nestos catchment.

3.3. Hydrologic Model Outputs

Altogether, the simulation results demonstrated a satisfactory correlation among the observed discharges and the simulated ones, as depicted in Figure 6. In terms of streamflow seasonality, the simulated discharges followed the Mediterranean area's pattern, i.e., high flows during late winter and spring, and low flows during summer and autumn. In the case of ERAI_50 and ERAI_25, the average discharges over the designated period from 1981 to 1995 were 20.7 m³/s and 21.9 m³/s, respectively. The aforementioned discharges were relatively higher than the 18.1 m³/s that was observed during the same period. This means that, in the case of ERAI_50 and ERAI_25, there was an overestimation of the river discharges of approximately 14.3% and 20.9%, respectively. On the other hand, for the ERAI_12.5 simulation, the discharges were equal to 15.8 m³/s, i.e., there is an underestimation of approximately 13.7% of the runoff. Moreover, in all simulation runs, the observed dry period of 1989 to 1994 was clearly depicted, as well as the maximum monthly flows that occurred in the period 1986–1987.

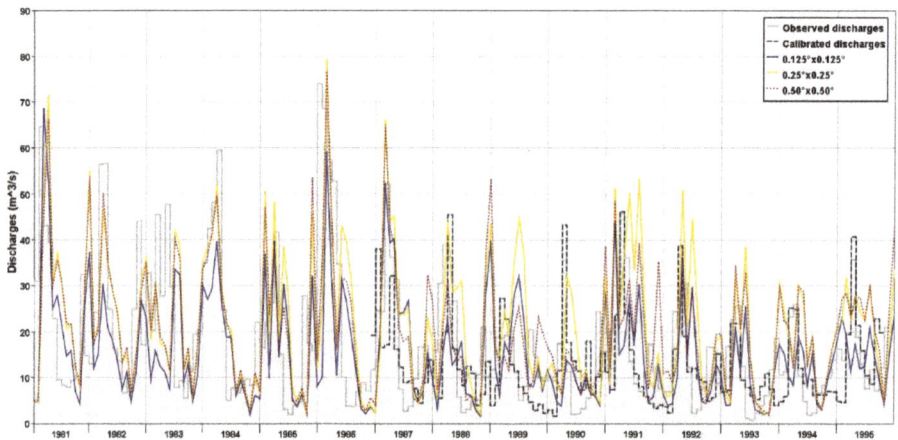

Figure 6. Observed (hashed area) and initial calibrated (black dotted curve) discharges versus simulated discharges of 0.125° × 0.125°, 0.25° × 0.25°, and 0.50° × 0.50° spatial analysis (blue, yellow, and brown curves, respectively).

Regarding the observed discharges at the interannual (15 years of data) time scale, the maxima of 34.8 m³/s and 35.7 m³/s occurred during April and May, as demonstrated in Figure 7 for the MAM (March–April–May) time period. The simulated discharges of 31.1 m³/s at 0.50° × 0.50°, 33.7 m³/s at 0.25° × 0.25°, and 23.9 m³/s at 0.125° × 0.125° resolution are shown for the same period. As for the minimum discharges, all scenarios matched the observed minimum flows in autumn. The observed minimum of 5.5 m³/s in September was slightly lower than the 7.7 m³/s and 6.5 m³/s of the ERAI_50 and ERAI_25, as demonstrated in Figure 7 for the SON (September–October–November) time period. For the ERAI_12.5 simulations, the minimum of 5.6 m³/s occurred in October. The relatively high streamflows of June, as shown in the JJA (June–July–August) diagram of Figure 7, were attributed to increased precipitation that generally occurs during the beginning of the summer season.

The coefficient of determination R^2 for the simulations was equal to 0.63 for the 0.125° × 0.125°, 0.69 for the 0.25° × 0.25°, and 0.66 for the 0.50° × 0.50° simulation. At the same time, the PBIAS for the same sequence of simulations was 24.81, −8.82, and −3.31, respectively. The monthly analysis of the results (Figure 8) revealed an increased correlation of all the datasets. In particular, the observed minimum flow of 0.9 m³/s was of the same magnitude as the 1.8 m³/s (ERAI_50), 1.9 m³/s (ERAI_25), and 1.6 m³/s (ERAI_12.5) simulations. At the same time, the maximum discharge of 74.0 m³/s was comparable to 82.8 m³/s (ERAI_50), 79.3 m³/s (ERAI_25), and 68.7 m³/s (ERAI_12.5). The median of all the datasets ranged between 12.5 m³/s and 19.0 m³/s, with the minimum and maximum values related to 0.125° × 0.125° and 0.25° × 0.25° simulations, respectively. Finally, what can also be observed is that the interquartile range, i.e., the area between the upper and lower quartiles, of the ERAI_50 and ERAI_25 datasets was almost identical.

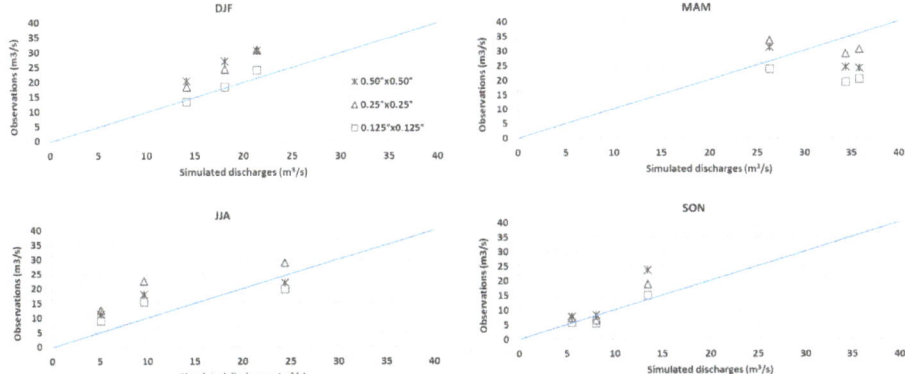

Figure 7. Quarterly correlation of observed and simulated discharges at the political borders of the transboundary Mesta/Nestos river basin.

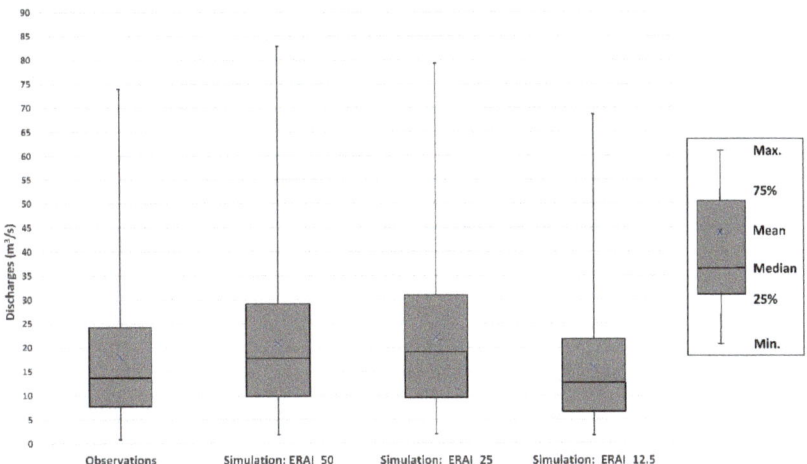

Figure 8. Boxplots comparing the Mesta/Nestos river discharges, as derived from the hydrologic simulation of the basin under different input gridded datasets.

4. Discussion

The aim of the present research was to assess the spatial resolution of reanalysis climatic data below or over which the hydrologic modeling simulation was not affected by the scale of the inputs. To address this issue, three different reanalysis datasets of 0.50° × 0.50°, 0.25° × 0.25°, and 0.125° × 0.125° resolution were used as inputs to a spatially distributed hydrologic model in order to generate discharges of a study basin.

The literature demonstrates a number of researches where either reanalysis products were evaluated with ground truth data [23,25,26] or reanalysis products of different origin were used for hydrologic simulations. Regarding the latter, Essou et al. [22] investigated the hydrologic response of 370 watersheds based on three reanalysis products with the same spatial resolution. The watershed areas ranged between 104 and 10,325 km², and they were allocated in five different climatic regions, including the Mediterranean and continental climates, which is the climate of the region to which the Mesta/Nestos basin belongs. The novelty of the current research consists of the assessment of reanalysis data coming from the same source, i.e., the ECMWF reanalysis product, but with different spatial resolutions, determining the performance of hydrologic simulations at a river basin scale. To the authors' knowledge, no similar analysis regarding the impact of the spatial analysis of ERA reanalysis gridded data on the hydrologic modeling procedure is referenced in the literature. At the same time, apart from the sensitivity analysis of the simulated discharges for the ERAI_50, ERAI_25, and ERAI_12.5 datasets, the current research explicitly provides a comparative analysis of the aforementioned reanalysis gridded variables of precipitation and temperature, and important outputs are designated.

As for the gridded variables of temperature and precipitation, the performed analysis demonstrated that, in the case of temperatures, the lower temperatures originated from the data with the finer spatial resolution. On the other hand, the higher temperatures over the basin area were attributed to the data with coarser spatial analysis. This difference was observed as the higher-resolution datasets managed to detect the altitude and the vertical temperature gradient in a more detailed way. The representation of the terrain's elevation in the case of finer-resolution datasets attributes the regional climatic characteristics in a more accurate manner, particularly in mountainous watersheds. In the case of a coarser mesh, these topographic variations are smoothed due to the average elevation that is attributed to each cell. Erum et al. [31] used three high-resolution gridded datasets with different resolutions, namely, NARR, ANUSPLIN, and CaPA, with resolutions of ~32 km, 10 km, and 15 km, respectively, for a case study area in western Canada. The coarsest of the three aforementioned datasets demonstrated warmer temperatures in all seasons except for winter, where the second in terms of resolution data presented almost the same outputs as the first. Regarding the precipitation variable, the analysis revealed negligible statistical differences in the two parts of the basin. The Mesta/Nestos catchment, due to its complex topography and adjacency to the north Aegean Sea, probably required much more than basic elevation information for the successful climate interpolation of precipitation data. This finding is consistent with that of Daly et al. [33] and Leung and Qian [50], who proved that elements such as elevation, location, the vicinity of the sea, catchment topographic orientation, and vertical atmospheric layer are essential for the interpolation of precipitation.

The catchment size seems to affect the impact of resolution of precipitation input on stream discharge, since the impact of fine-resolution precipitation input was negligible for the specific catchment. In small catchment areas up to 100.0 km^2, Berne et al. [51] demonstrated that, under specific characteristics, such as a slope between 1% and 10%, impermeability between 10% and 60%, and the Mediterranean climate, the required resolution of precipitation input should be 5.2 km. For catchment areas larger than 1000 km^2, it was demonstrated [52] that they are not highly affected by the spatial resolution of precipitation. According to Fu et al. [52], all simulations that are related to lower discharges are linked with datasets of finer resolution, apart from during the warm periods of the year. Temperature variability seems to be a primary factor of discharge reduction at the lower resolution, while rainfall coverage can remain a principal factor at finer resolutions. Kouwen [53] also showed in his research that lower-resolution radar rainfall data of 10.0 km × 10.0 km can sufficiently be used for modeling floods compared to higher-resolution data of 2.0 km × 2.0 km.

The increased resolution has the benefit of reducing numerical truncation errors [54], while it permits the simulation of fine-scale details. However, in complex terrains, such as mountainous areas coupled with plains, and in complex climates, such as continental climate effected by seacoast climate, lower-resolution data might be more suitable for the simulation and prediction of the temperature [55]. Colle et al. [56] reached similar conclusions by indicating that higher resolution is more sensitive to the convective parameterization and reduces the accuracy of climate parameters, especially when the topography is complex. As demonstrated in the results, in the case of low and medium resolution, the overestimation of the discharges was on the order of 14.3% and 20.9%, respectively, while, in the case of higher resolution, there was an underestimation of 13.7%. Similar hydrologic modeling performance was presented in a German catchment of approximately 4000 km^2 [57]. That research, where the spatial variation was based on kriging from point measurements, proved that simulations with coarser-resolution data outperformed the finer-resolution simulations. On the other hand,

Haddeland et al. [58] concluded that, when the topographic and land-cover data resolution is altered, as well as the forcing variables, then the coarser-resolution forced simulations are more biased than finer-resolution simulations. In any case, since the runoff response of a watershed is not directly proportional to precipitation, but it is also governed by the physical characteristics of a watershed (e.g., topography, soil, and land cover), differences in the timing and volume of the produced discharges is bound to occur. The need for amelioration of the distributed model structure to make better use of inputs of fine resolution could be proposed as a solution to the simulation process uncertainty. The later output agrees with Bell and Moore [59], who demonstrated in their research, contrary to what they initially believed, that, in the case of gridded rainfalls, the best performance is obtained when lower-resolution radar data are utilized, and they suggested a revision of the model structure.

An additional reason behind the simulated outputs is the dependence of the river's flow, particularly in the summer period, on the snowmelt process. The temperature is the principal variable that affects the melting of snow when using a snowmelt runoff model. For the current case basin, the overestimated discharges of ERAI_50 and EAI_25 data coincided with the high temperatures that were presented in both datasets. Thus, it is believed that the ERA dataset resolution does not compensate for errors in the model calibration in the region. However, the application of different hydrological models that contain snowmelt-computing options could provide different outputs, since each model has its own snowmelt algorithm. Verden et al. [60] investigated the performance of snowmelt algorithms that are integrated into watershed models, namely, HEC-1, SSARR UBC, NWSRFS, PRMS, SHE, SRM, and TANK, and they reached the conclusion that different snowpack melt quantities were derived from each model in the same case area. Jain et al. [61] demonstrated that warmer climatic conditions increase the annual stream flow but not severely. At the same time, Lutz et al. [62] concluded that, in the case of climate change, i.e., increased temperatures, the snowmelt process is caused by an increase in precipitation and from accelerated melt due to the higher temperatures. The aforementioned finding might be contradictory to the temperature's impact on evapotranspiration, meaning that high temperatures result in increased evapotranspiration and, hence, decreased water flows. However, Koedyk and Kingston [63], in order to investigate the influence of PET on a river's runoff, used six different PET methods for a temperature increase of 2.0 °C as derived from five general circulation models. The output of their research demonstrated that, in rivers of continuous flow, the impact is relatively small, i.e., under 5% at the monthly scale and at most 5.2% at the annual scale.

Concerning the accuracy of the simulated discharges, their statistical assessment demonstrated the reliability of the outputs, i.e., $R^2 > 0.63$ for each of the simulated datasets, since values greater than 0.5 are typically considered acceptable [64]. It should be mentioned that the derived model evaluation statistics are sensitive to high values (outliers) and insensitive to additive and proportional differences between model predictions and measured data [65]. However, the daily and monthly datasets used in this study do not fall into the previous categories. As for the PBIAS method, low-magnitude values indicate accurate model simulation, as in the case of the 0.50° × 0.50° and 0.25° × 0.25° resolution, while positive and negative values indicate model underestimation and overestimation bias, respectively [48]. Moreover, according to the general performance ratings for recommended statistics for a monthly time step [65], a PBIAS (%) of the streamflow with ranges of ±10 < PBIAS < ±15 and ±15 < PBIAS < ±25 is considered good and satisfactory, respectively. In the proposed research, the results derived from climate data of 0.50° × 0.50° and 0.25° × 0.25° (PBIAS equal to −3.31 and −8.82, respectively) were classified as good, and those of 0.125° × 0.125° (PBIAS equal to +24.2) were classified as satisfactory [66].

Considering the number of areas in Europe, as well as other parts of the world, where dense meteorological monitoring networks are lacking, the potential offered by the gridded and reanalysis datasets is unique. The proposed research outputs are considered appropriate for assisting the selection of reanalysis data resolution, mostly in cases with similar geomorphological characteristics and climatic conditions. Moreover, since the climatic change impact on water resources is ongoing research, it is believed that the present study could offer added value to the selection process of reanalysis data that will be used for the bias correction of climate change datasets.

5. Conclusions

Data availability and, in particular, precipitation and temperature data series are crucial for the hydrologic simulation of river basins. Apart from the accuracy and reliability of data, their spatial coverage and density have increased impact on the hydrologic modeling behavior. Currently, the availability of gridded dataset products of various resolutions provides solutions in areas of coarse gauge networks or even in regions with a lack of observations. However, this plethora of data sources that provide meteorological and climatological variables at meshes of varying resolution could be a bottleneck in hydrologic modeling.

The assessment of the impact of three different ERA-Interim reanalysis datasets, in terms of spatial resolution, on river basin hydrology suggests that, for the runoff simulations at a daily time step, the most appropriate dataset is of medium resolution. The produced biases in the case of climatic variables of coarser or more refined resolution are relatively low ($\leq \pm 10\%$); thus, these data could also be used for the long-term management of water resources. An important factor of the outputs is the dependence of the summer runoff on the snowmelt process. Moreover, the scale of the basin plays an important role in the selection of the most appropriate resolution. Overall, the results presented that reanalysis data sources could be used as proxies to successfully force hydrological models. Finally, the proposed research could also shed light in studies focusing on climate change impacts on hydrology; thus, the question whether climate input data having higher spatial resolution result in better model simulations could be explored.

Author Contributions: In this research, all authors equally contributed to the conceptualization of the paper, the simulations, and the interpretation of the outputs, as well as to the writing, review, and editing of the paper. All authors have read and agreed to the published version of the manuscript.

Funding: This research received no external funding.

Conflicts of Interest: The authors declare no conflicts of interest.

References

1. Brigode, P.; Oudin, L.; Perrin, C. Hydrological model parameter instability: A source of additional uncertainty in estimating the hydrological impacts of climate change? *J. Hydrol.* **2013**, *476*, 410–425. [CrossRef]
2. Refsgaard, J.C.; Storm, B. Construction, Calibration and Validation of Hydrological Models. In *Distributed Hydrological Modelling. Water Science and Technology Library*; Abbott, M.B., Refsgaard, J.C., Eds.; Springer: Dordrecht, Netherlands, 1990; Volume 22.
3. O'Riordan, T. *Environmental Science for Environmental Management*; Routledge: London, UK, 2014.
4. Chen, Y.C.; Wei, C.; Yeh, H.C. Rainfall network design using kriging and entropy. *Hydrol. Process.* **2008**, *22*, 340–346. [CrossRef]
5. Xu, H.; Xu, C.Y.; Chen, H.; Zhang, Z.; Li, L. Assessing the influence of rain gauge density and distribution on hydrological model performance in a humid region of China. *J. Hydrol.* **2013**, *505*, 1–12. [CrossRef]
6. Anctil, F.; Lauzon, N.; Andréassian, V.; Oudin, L.; Perrin, C. Improvement of rainfall–runoff forecasts through mean areal rainfall optimization. *J. Hydrol.* **2006**, *328*, 717–725. [CrossRef]
7. Woods, R.A.; Sivapalan, M.; Duncan, M. Investigating the representative elementary area concept: An approach based on field data. *Hydrol. Process.* **1995**, *6*, 291–312. [CrossRef]
8. Bathurst, J.C. Physically based distributed modeling of upland catchment using the Systeme Hydrologique Europeen. *J. Hydrol.* **1986**, *87*, 79–123. [CrossRef]
9. World Meteorological Organization. Density of stations for a network. In *Hydrology—From Measurement to Hydrological Information, Vol 1, Guide to Hydrological Practices*, 6th ed.; WMO-168; World Meteorological Organization: Geneva, Switzerland, 2008.
10. Khan, A.J.; Koch, M.; Chinchilla, K.M. Evaluation of Gridded MultiSatellite Precipitation Estimation (TRMM-3B42-V7) Performance in the Upper Indus Basin (UIB). *Climate* **2018**, *6*, 76. [CrossRef]
11. Rossi, M.; Kirschbaum, D.; Valigi, D.; Mondini, A.; Guzzetti, F. Comparison of Satellite Rainfall Estimates and Rain Gauge Measurements in Italy, and Impact on Landslide Modeling. *Climate* **2017**, *5*, 90. [CrossRef]

12. Haylock, M.R.; Hofstra, N.; Klein Tank, A.M.G.; Klok, E.J.; Jones, P.D.; New, M. A European daily high-resolution gridded data set of surface temperature and precipitation for 1950–2006. *J. Geophys. Res.* **2008**, *113*. [CrossRef]
13. Lazoglou, G.; Anagnostopoulou, C.; Skoulikaris, C.; Tolika, K. Bias Correction of Climate Model's Precipitation Using the Copula Method and Its Application in River Basin Simulation. *Water* **2019**, *11*, 600. [CrossRef]
14. Nerini, D.; Zulkafli, Z.; Wang, L.; Onof, C.; Buytaert, W.; Lavado-Casimiro, W.; Guyot, J. A Comparative Analysis of TRMM–Rain Gauge Data Merging Techniques at the Daily Time Scale for Distributed Rainfall–Runoff Modeling Applications. *J. Hydrometeorol.* **2015**, *16*, 2153–2168. [CrossRef]
15. Nastos, P.T.; Kapsomenakis, J.; Douvis, K.C. Analysis of precipitation extremes based on satellite and high-resolution gridded data set over Mediterranean basin. *Atmos. Res.* **2013**, *131*, 46–59. [CrossRef]
16. Adam, J.C.; Lettenmaier, D.P. Adjustment of global gridded precipitation for systematic bias. *J. Geophys. Res. Atmos.* **2003**, *108*, 4257. [CrossRef]
17. Faiz, M.A.; Liu, D.; Fu, Q.; Sun, Q.; Li, M.; Baig, F.; Li, T.; Cui, S. How accurate are the performances of gridded precipitation data products over Northeast China? *Atmos. Res.* **2018**, *211*, 12–20. [CrossRef]
18. Hofstra, N.; Haylock, M.; New, M.; Jones, P.D. Testing E-OBS European high-resolution gridded data set of daily precipitation and surface temperature. *J. Geophys. Res.* **2009**, *114*. [CrossRef]
19. Gao, L.; Schulz, K.; Bernhardt, M. Statistical downscaling of ERA-interim forecast precipitation data in complex terrain using lasso algorithm. *Adv. Meteorol.* **2014**, *2014*. [CrossRef]
20. Jahani, B.; Mohammadi, B. A comparison between the application of empirical and ANN methods for estimation of daily global solar radiation in Iran. *Theor. Appl. Climatol.* **2019**, *137*, 1257. [CrossRef]
21. Fuka, D.R.; Walter, M.T.; MacAlister, C.; Degaetano, A.T.; Steenhuis, T.S.; Easton, Z.M. Using the Climate Forecast System Reanalysis as weather input data for watershed models. *Hydrol. Process.* **2014**, *28*, 5613–5623.
22. Essou, G.R.C.; Sabarly, F.; Lucas-Picher, P.; Brissette, F.; Poulin, A. Can precipitation and temperature from meteorological reanalyses be used for hydrological modeling? *J. Hydrometeorol.* **2016**. [CrossRef]
23. Dee, D.; Uppala, S.M.; Simmons, A.J.; Berrisford, P.; Poli, P.; Kobayashi, S.; Andrae, U.; Balmaseda, M.A.; Balsamo, G.; Bauer, P.; et al. The ERA-Interim reanalysis: Configuration and performance of the data assimilation system. *Quart. J. R. Meteor. Soc.* **2011**, *137*, 553–597. [CrossRef]
24. Bosilovich, M.G. Regional climate and variability of NASA MERRA and recent reanalyses: U.S. summertime precipitation and temperature. *J. Appl. Meteor. Climatol.* **2013**, *52*, 1939–1951. [CrossRef]
25. Lorenz, C.; Kunstmann, H. The hydrological cycle in three state-of-the-art reanalyses: Intercomparison and performance analysis. *J. Hydrometeorol.* **2012**, *13*, 1397–1420. [CrossRef]
26. Grusson, Y.; Anctil, F.; Sauvage, S.; Sánchez Pérez, J.M. Testing the SWAT Model with Gridded Weather Data of Different Spatial Resolutions. *Water* **2017**, *9*, 54. [CrossRef]
27. Gao, L.; Bernhardt, M.; Schulz, K. Elevation correction of ERA-Interim temperature data in complex terrain. *Hydrol. Earth Syst. Sci.* **2012**, *16*, 4661–4673. [CrossRef]
28. Muppa, S.K.; Anandan, V.K.; Kesarkar, K.A.; Rao, S.V.B.; Reddy, P.N. Study on deep inland penetration of sea breeze over complex terrain in the tropics. *Atmospheric. Res.* **2012**, *104*, 209–216. [CrossRef]
29. Hu, Z.; Hu, Q.; Zhang, C.; Chen, X.; Li, Q. Evaluation of reanalysis, spatially interpolated and satellite remotely sensed precipitation data sets in central Asia. *J. Geophys. Res. Atmos.* **2016**, *121*, 5648–5663. [CrossRef]
30. Donat, M.G.; Sillmann, J.; Wild, S.; Alexander, L.V.; Lippmann, T.; Zwiers, F.W. Consistency of Temperature and Precipitation Extremes across Various Global Gridded In Situ and Reanalysis Datasets. *J. Clim.* **2014**, *27*, 5019–5035. [CrossRef]
31. Eum, H.I.; Dibike, Y.; Prowse, T.; Bonsal, B. Inter-comparison of high-resolution gridded climate data sets and their implication on hydrological model simulation over the Athabasca Watershed, Canada. *Hydrol. Process.* **2014**, *28*, 4250–4271. [CrossRef]
32. Lundquist, J.D.; Minder, J.R.; Neiman, P.J.; Sukovich, E. Relationships between Barrier Jet Heights, Orographic Precipitation Gradients, and Streamflow in the Northern Sierra Nevada. *J. Hydrometeorol.* **2010**, *11*, 1141–1156. [CrossRef]
33. Daly, C.; Halbleib, M.; Smith, J.I.; Gibson, W.P.; Doggett, M.K.; Taylor, G.H.; Curtis, J.; Pasteris, P.P. Physiographically sensitive mapping of climatological temperature and precipitation across the conterminous United States. *Int. J. Climatol. A J. R. Meteorol. Soc.* **2011**, *28*, 2031–2064. [CrossRef]

34. Woo, M.K.; Thorne, R. Snowmelt contribution to discharge from a large mountainous catchment in subarctic Canada. *Hydrol. Process.* **2006**, *20*, 2129–2139. [CrossRef]
35. Bhattacharya, T.; Khare, D.; Arora, M. A case study for the assessment of the suitability of gridded reanalysis weather data for hydrological simulation in Beas river basin of North Western Himalaya. *Appl. Water Sci.* **2019**, *9*, 110. [CrossRef]
36. Tarek, M.; Brissette, F.P.; Arsenault, R. Evaluation of the ERA5 reanalysis as a potential reference dataset for hydrological modeling over North-America. *Hydrol. Earth Syst. Sci. Discuss.* **2019**, [CrossRef]
37. Skoulikaris, C.H.; Ganoulis, J. Climate Change Impacts on River Catchment Hydrology Using Dynamic Downscaling of Global Climate Models. In *National Security and Human Health Implications of Climate Change, NATO Science for Peace and Security Series C: Environmental Security*; Fernando, H., Klaić, Z., McCulley, J., Eds.; Springer: Berlin, Germany, 2012; pp. 281–287.
38. Bank of Greece. *Environmental, Economic and Social Impacts Due to Climate Change in Greece*; Bank of Greece: Athens, Greece, 2011; p. 546.
39. UNECE. *Second Assessment of Transboundary Rivers, Lakes and Groundwaters*; UNECE: Geneva, Switzerland, 2011.
40. Balsamo, G.; Albergel, C.; Beljaars, A.; Boussetta, S.; Brun, E.; Cloke, H.; De Rosnay, P. ERA-Interim/Land: A global land surface reanalysis data set. *Hydrol. Earth Syst. Sci.* **2015**, *19*, 389–407. [CrossRef]
41. Ledoux, E.; Girard, G.; de Marsily, G.; Deschenes, J. Spatially distributed modelling: Conceptual approach, coupling surface water and ground water. In *Unsaturated Flow Hydrologic Modelling-Theory and Practice*; Morel-Seytoux, H.J., Ed.; NATO ASI Series S 275; Kluwer Academic: Boston, CA, USA, 1989; pp. 435–454.
42. Violette, S.; Ledoux, E.; Goblet, P.; Carbonnel, J.-P. Hydrologic and thermal modeling of an active volcano: The Piton de la Fournaise, Reunion. *J. Hydrol.* **1997**, *191*, 37–63. [CrossRef]
43. Artinyan, E.; Habets, F.; Noilhan, J.; Ledoux, E.; Dimitrov, D.; Martin, E.; Le Moigne, P. Modelling the water budget and the riverflows of the Maritsa basin in Bulgaria. *Hydrol. Earth Syst. Sci.* **2008**, *12*, 21–37. [CrossRef]
44. Natural Resources Conservation Service (NRCS). *National Engineering Handbook, Part 630 Hydrology*; U.S. Department of Agriculture: Washington, DC, UAS, 1993; Chapt. 11, Snowmelt.
45. Rango, A.; Martinec, J. Revisiting the degree-day method for snowmelt computations. *J. Am. Water Resour. Assoc.* **1995**, *31*, 657–669. [CrossRef]
46. Etchevers, P.; Golaz, C.; Habets, F. Simulation of the water budget and the river flows of the Rhone basin from 1981 to 1994. *J. Hydrol.* **2001**, *244*, 60–85. [CrossRef]
47. Etchevers, P.; Martin, E. Impact d'un changement climatique sur le manteau neigeux et l'hydrologie des bassins versants de montagne. In Proceedings of the Colloque International «L'eau en montagne», Mégève, France, 6 September 2002.
48. Gupta, H.V.; Sorooshian, S.; Yapo, P.O. Status of automatic calibration for hydrologic models: Comparison with multilevel expert calibration. *J. Hydrologic. Eng.* **1999**, *4*, 135–143. [CrossRef]
49. Jolliffe, I.T.; Cadima, J. Principal component analysis: A review and recent developments. *Phil. Trans. R. Soc. A* **2016**, *374*, 20150202. [CrossRef]
50. Leung, L.R.; Qian, Y. The sensitivity of precipitation and snowpack simulations to model resolution via nesting in regions of complex terrain. *J. Hydrometeorol.* **2003**, *4*, 1025–1043. [CrossRef]
51. Berne, A.; Delrieu, G.; Creutin, J.D.; Obled, C. Temporal and spatial resolution of rainfall measurements required for urban hydrology. *J Hydrol.* **2004**, *299*, 166–179. [CrossRef]
52. Fu, S.; Sonnenborg, T.O.; Jensen, K.H.; He, X. Impact of Precipitation Spatial Resolution on the Hydrological Response of an Integrated Distributed Water Resources Model. *Vadose Zone J.* **2011**, *10*, 25–36. [CrossRef]
53. Kouwen, N. *SIMPLE—A Watershed Model for Flood Forecasting. Users' Manual*; Department of Civil Engineering, University of Waterloo: Waterloo, ON, Canada, 1986; p. 130.
54. Wang, X.; Steinle, P.; Seed, A.; Xiao, Y. The Sensitivity of Heavy Precipitation to Horizontal Resolution, Domain Size, and Rain Rate Assimilation: Case Studies with a Convection-Permitting Model. *Adv. Meteorol.* **2016**, *2016*. [CrossRef]
55. Giunta, G.; Salerno, R.; Ceppi, A.; Ercolani, G.; Mancini, M. Effects of Model Horizontal Grid Resolution on Short- and Medium-Term Daily Temperature Forecasts for Energy Consumption Application in European Cities. *Adv. Meteorol.* **2019**, *2019*. [CrossRef]
56. Colle, B.A.; Olson, J.B.; Tongue, J.S. Multiseason verification of the MM5. Part II: Evaluation of high-resolution precipitation forecasts over the northeastern United States. *Weather. Forecast.* **2003**, *18*, 458–480. [CrossRef]

57. Das, T.; Bárdossy, A.; Zehe, E.; He, Y. Comparison of conceptual model performance using different representations of spatial variability. *J. Hydrol.* **2008**, *356*, 106–118. [CrossRef]
58. Haddeland, I.; Matheussen, B.V.; Lettenmaier, D.P. Influence of spatial resolution on simulated streamflow in a macroscale hydrologic model. *Water Resour. Res.* **2002**, *38*, 291–2910. [CrossRef]
59. Bell, V.A.; Moore, R.J. The sensitivity of catchment runoff models to rainfall data at different spatial scales. *Hydrol. Earth Syst. Sci.* **2000**, *4*, 653–667. [CrossRef]
60. Verdhen, A.; Chahar, B.R.; Sharma, O.P. Snowmelt modelling approaches in watershed models: Computation and comparison of efficiencies under varying climatic conditions. *Water Resour. Manag.* **2014**, *28*, 3439–3453. [CrossRef]
61. Jain, S.K.; Goswami, A.; Saraf, A.K. Assessment of Snowmelt Runoff Using Remote Sensing and Effect of Climate Change on Runoff. *Water Resour. Manag.* **2010**, *24*, 1763. [CrossRef]
62. Lutz, A.F.; Immerzeel, W.W.; Shrestha, A.B.; Bierkens, M.F.P. Consistent increase in High Asia's runoff due to increasing glacier melt and precipitation. *Nat. Clim. Chang.* **2014**, *4*, 587–592. [CrossRef]
63. Koedyk, L.P.; Kingston, D.G. Potential evapotranspiration method influence on climate change impacts on river flow: A mid-latitude case study. *Hydrol. Res.* **2016**, *47*, 951–963. [CrossRef]
64. Santhi, C.; Arnold, J.G.; Williams, J.R.; Dugas, W.A.; Srinivasan, R.; Hauck, L.M. Validation of the SWAT model on a large river basin with point and nonpoint sources. *J. Am. Water Resour. Assoc.* **2001**, *37*, 1169–1188. [CrossRef]
65. Legates, D.R.; McCabe, G.J. Evaluating the use of "goodness-of-fit" measures in hydrologic and hydroclimatic model validation. *Water Resour. Res.* **1999**, *35*, 233–241. [CrossRef]
66. Moriasi, D.N.; Arnold, J.G.; Van Liew, M.W.; Bingner, R.L.; Harmel, R.D.; Veith, T.L. Model evaluation guidelines for systematic quantification of accuracy in watershed simulations. *Trans. ASABE* **2007**, *50*, 885–900. [CrossRef]

© 2019 by the authors. Licensee MDPI, Basel, Switzerland. This article is an open access article distributed under the terms and conditions of the Creative Commons Attribution (CC BY) license (http://creativecommons.org/licenses/by/4.0/).

Article

The Global Trend of the Net Irrigation Water Requirement of Maize from 1960 to 2050

Abdoulaye Oumarou Abdoulaye [1], Haishen Lu [1,*], Yonghua Zhu [1], Yousef Alhaj Hamoud [2] and Mohamed Sheteiwy [3]

1. State Key Laboratory of Hydrology-Water Resources and Hydraulic Engineering, College of Hydrology and Water Resources, Hohai University, Nanjing 210098, China; d2016002@hhu.edu.cn (A.O.A.); zhuyonghua@hhu.edu.cn (Y.Z.)
2. College of Agriculture Engineering, Hohai University, Nanjing 210098, China; yousef-hamoud11@hotmail.com
3. Department of Agronomy, Faculty of Agriculture, Mansoura University, Mansoura 35516, Egypt; salahco_2010@yahoo.com
* Correspondence: haishenlu@gmail.com

Received: 1 July 2019; Accepted: 14 October 2019; Published: 22 October 2019

Abstract: Irrigated production around the world has significantly increased over the last decade. However, climate change is a new threat that could seriously aggravate the irrigation water supplies and request. In this study, the data is derived from the IPCC Fifth Assessment Report (AR5). For the climate change scenarios, five Global Climate Models (GCMs) have been used. By using the CROPWAT approach of Smith, the net irrigation water requirement (IRnet) was calculated. For the estimation of the potential evapotranspiration (Epot), the method in Raziei and Pereira was used. According to representative concentration pathway (RCP) 4.5, these increases vary between 0.74% (North America) and 20.92% (North America) while the RCP 8.5 predict increases of 4.06% (sub-Saharan Africa) to more than 68% (North America). The results also show that the region of Latin America is the region with the large amount of IRnet with coprime value between 1.39 km^3/yr (GFDL 4.5) and 1.48 km^3/yr (CSIRO 4.5) while sub-Saharan Africa has the smallest IRnet amount between 0.13 km^3/yr (GFDL 8.5) and 0.14 km^3/yr (ECHAM 8.5). However, the most affected countries by this impact are those in sub-Saharan Africa. This study will probably help decision-makers to make corrections in making their decision.

Keywords: global; climate change; temperature; precipitation; Net Irrigation Water Requirement; maize

1. Introduction

The increasing speed of climate change and related changes in rainfall have officially influenced biological systems and biodiversity on Earth [1]. Indeed, agriculture and environmental change are deeply intertwined. Ecological changes are now having an impact on the agri-food sector, with implications that have unevenly adapted to the world. Future environmental changes are likely to harm crop production in low-perch countries, while the consequences in the Nordic are likely to be less affected. Besides, environmental changes will likely create a danger of food insecurity for some powerless gatherings, such as the poor. The World Health Organization (WHO) assessments over the last 30 years have already helped more than 150,000 victims per year due to global warming and precipitation trends caused by anthropogenic climate change [2]. Many other studies have been undertaken to examine the impact of climate change on animal and plants health [3–8]. According to some studies on a global scale, regional variations associated with climate change are not expected to lead to significant changes in food production over the next century [5,9]. Several simulation models have been made about climate change [10]. The majority of climate simulation models show: an increase in the global temperature average; the temperatures are projected to increase between 1.4 and 5.88 °C by the end of this century; a related rise in sea level is also expected [11]. The impact on the economy was also studied [12–16], with developing countries expected to suffer most of the damage caused by climate change, while the wealthiest countries are likely to be less affected [17–19].

Precise and multi-variable predictions of anthropogenic climate change are needed to assess the effects [20]. Several studies have addressed the problem of the impact of climate change on agriculture. Several aspects of the question were analyzed. One of the topics is the impact of climate change on irrigation [21–24]. Among the sectors of water use, irrigation will be most influenced by the effects of climate change [25–31]. Agriculture and mainly irrigated agriculture is the sector with by far the primary consumptive water use and water withdrawal. To assess the gravity of irrigation on the available water resources, an estimate has been made mutually for the irrigation water requirement and also, irrigation water withdrawal [32]. Irrigation ensures summer production and produces valuable crops that would otherwise be impossible to cultivate [33,34].

Nevertheless, the effects on regional and local food supplies in some low latitude regions could amount to significant percentage changes in current production [35]. Asia's intensive agriculture consumes 20% of its internal renewable resources, of which more than 80% goes to irrigation. In most low-rainfall areas of the Middle East, North Africa, and Central Asia, most of the exploitable water is already very scarce, with 80–90% of this water destined for agriculture. Rivers and aquifers are, therefore, operating beyond their sustainable levels. In the long list of possible problems from worldwide warming, the menaces to world agriculture stand out as among the most important [36]. Regional predictions are needed for improving the assessments of vulnerability to and impacts of change [37].

Several recent articles have focused on the impact of climate change on plant production [38–40], but also on how to adapt to climate change [41–44]. In 2002, Petra Doll and Stefan Siebert presented a global model on the demand for irrigation water. The model simulates the cropping patterns, the growing seasons, and the net and gross irrigation requirements, distinguishing between two crops, rice and non-rice [45,46]. The correspondence between their model results and independent assessments of irrigation water use is considered to be adequate for applying the model in global and continental studies.

In this study, the variations in long-term averages of precipitation and temperature are the only characteristics that define climate change. In our agro-climatological approach, one cereal was selected and it was assumed that it was planted under optimal conditions for their growth. Any change in the dimension and location of the flooded areas due to an adaptation to climate change or for any other reason were neglected. In addition, the increase in CO_2 and its direct effects crops had to be ignored due to insufficient quantitative knowledge.

Therefore, based on IPCC-5 data, the purpose of this study was to: First, to estimate the variation of IRnet of maize in the past and the future around the world; second, to map this variation. At the end of this study, it is possible to observe the evolution of the IRnet under the effect of climate change according to the models.

2. Materials and Methods

2.1. Data

In this study, climate change data were derived from the IPCC Fifth Assessment Report (AR5) (http://www.ipcc-data.org/sim/gcm_global/index.html). Climate change scenarios were utlized from five GCMs (CSIRO, ECHAM.MPI-ESM-LR, GFDL.ESM2G, MIROC5, and NCAR.CCSM4), moreover for each model two representative concentration pathways (RCP) were chosen. They were RCP 4.5 and 8.5. The global map of the currently irrigated areas has been uploaded to the food and agriculture organization (FAO) website (http://www.fao.org/nr/water/aquastat/irrigationmap/index10.stm). This map represents the totality of the growing regions irrigated in 2005.

2.2. Net Irrigation Requirement Model

Following the CROPWAT approach of Smith [47], the net irrigation requirement per unit irrigated area through the growing season is calculated, with a daily time step, as the variance between the effective precipitation and the crop-specific potential evapotranspiration as:

$$IRnet = kc.Epot - P_{eff} \quad if \quad kc.Epot > P_{eff} \tag{1}$$

$$IRnet = 0$$

Otherwise, with $IRnet$ instead of the net irrigation requirement per unit area [mm/d]; P_{eff} is the effective precipitation [mm/d]; $Epot$ is the potential evapotranspiration [mm/d]; kc is the crop coefficient.

The crop coefficient, kc, depends on the crop type (maize) and the day of the growing period. P_{eff} is the fraction of P (the total precipitation) that is accessible to the crop and does not run off. Without detailed site-specific information, P_{eff} is very hard to determine. A simple estimate following the USDA soil conservation method is used, as cited in Smith [47], with:

$$P_{eff} = \frac{P(4.17 - 0.2P)}{4.17} \quad for \quad P < 8.2 \tag{2}$$

$$P_{eff} = 4.17 + 0.1P \quad for \quad P \geq 8.3 \tag{3}$$

CROPWAT uses monthly rainfall data from which 10-day-averages are derived as input for the calculations [47]. The application of Equation (1) with daily rainfall values, i.e., the days with and without precipitation, would lead to a gross overestimation of the net irrigation water requirement.

For the estimation of $Epot$, the method in Raziei and Pereira [48] was used. By replacing Ra by his equation:

$$Epot = 0.0135 \, k_{Rs} \frac{Ra}{\lambda} \sqrt{Tmax - Tmin}(T + 17.8) \tag{4}$$

$$Rs = k_{Rs} \sqrt{Tmax - Tmin}(T + 17.8) Ra \tag{5}$$

where:

$$Ra = \frac{Rs}{k_{Rs} \sqrt{Tmax - Tmin}} \tag{6}$$

In addition, assuming that $k_{Rs} = 0.17$ in Equations (4) and (6), Equation (4) gives after simplification:

$$Epot = 0.0135 \frac{Rs}{\lambda} (T + 17.8) \tag{7}$$

where Rs is the solar radiation (MJ/m²/day). It was obtained by adding the surface down-welling shortwave flow air and surface down-welling longwave flux in the air; λ is commonly equal to 2.45; T is the mean temperature (°C); Ra is extraterrestrial radiation (MJ/m²/d);

2.3. Climate Input

The temperature and precipitation data of each model was converted according to the units of the different formulas. The same method was applied for the solar radiation data, which was also calculated for each day from 1960 to 2050. After having calculated the P_{eff} from Equations (2) and (3), as well as the $IRnet$ (Equation (1)) and the potential evapotranspiration Equation (7) of the selected plant (according to the data of the five models), the results were used in ArcGIS to observe more closely the evolution of the net irrigation of the plant.

3. Results

According to the FAO global map of irrigation areas, the amounts of irrigation water (IRnet) vary from one model to another, but also from one RCP to another of the same model. The period 1960–1999 is considered here as being historical years. The year 1960 is the year of reference. The period 2050 is be regarded as the future, which allows appreciating the evolution of IRnet.

3.1. The Net Irrigation Water Requirements (IRnet) for Historical Period

3.1.1. The IRnet in 1960

The region of Latin America and the Caribbean has a total IRnet more than 1.34 km^3/year, followed by the East Asia and Pacific region with a quantity of 1 km^3/yr. The parts of the Middle East and North Africa with only 0.25 km^3/yr front the region of sub-Saharan Africa with 0.13 km^3/year (Table 1). Figure 1 shows the distribution of FAO areas in 2005. This study observed that areas using large quantities of *IRnet* are located between the latitude of Cancer and that of Capricorn, which are located respectively north and south of the equator. The quantities vary between 2000 and 2500 mm/year. The countries that lie beyond these two latitudes, there are quantities between 160 and 1800 mm^3/year.

Table 1. Historical regional net irrigation water requirement.

YEAR	Region	Surface (km^2)	Min (mm/yr)	Max (mm/yr)	Mean (mm/yr)	Change of Irnet (%)	Total of Irnet (km^3/yr)
1960	Latin America and Caribbean	673.68	710.13	2520.92	1992.63		1.34
	South Asia	327.56	583.75	2374.34	1835.98		0.60
	Sub-Saharan Africa	63.39	1431.73	2423.92	2028.61		0.13
	Europe and Central Asia	395.01	159.76	1716.05	873.67		0.34
	Middle East and North Africa	154.29	1018	2241.11	1613.53		0.25
	East Asia and Pacific	709.39	383.82	2370.29	1404.84		1.00
	North America	729.38	205.72	2082.77	1105.92		0.81
1999	Latin America and Caribbean	673.68	727.22	2570.09	2025.22	1.63	1.36
	South Asia	327.56	626.89	2390.15	1861.01	1.36	0.61
	Sub-Saharan Africa	63.39	1438.55	2453.89	2070.11	2.05	0.13
	Europe and Central Asia	395.01	173.77	1741.73	915.47	4.78	0.36
	Middle East and North Africa	154.29	1061.68	2285.73	1663.75	3.11	0.26
	East Asia and Pacific	709.39	399.29	2417.71	1441.97	2.64	1.02
	North America	729.38	246.46	2108.19	1135.32	2.66	0.83

Surface = total irrigation area according to the FAO in 2005, Min = the minimum value of the Irnet in the region, Max = maximum value of the Irnet in the region, Mean = the average of Irnet in the region, Variation of Irnet = comparison of Means compare to 1960, Total of Irnet = global irrigation water requirement used.

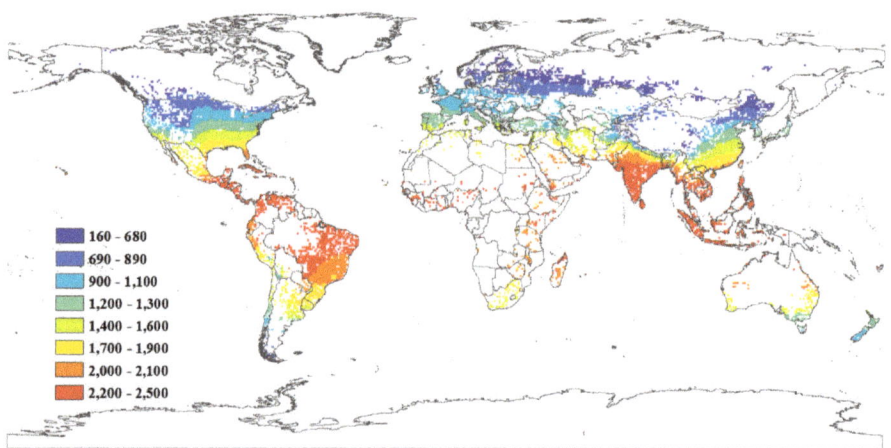

Figure 1. Map of the historical global net irrigation water requirement in 1960 of the area actually irrigated expressed as a percentage of the area equipped for irrigation in mm/yr.

3.1.2. The IRnet in 1999

Figure 2 shows the evolution of the IRnet in 1999. There is an increase in IRnet in a global manner. As a result, the region of Europe and Central Asia, for example, saw a rise of 4.78%, as did the East Asian and Pacific region with an increase of 2.64%, compared with 1960. However, the region of Latin America and the Caribbean has the highest IRnet total at 1.36 Km^3/year followed by the East and Pacific region (1.02 km^3/yr) and the region of North America (0.83 km^3/year) (Table 1).

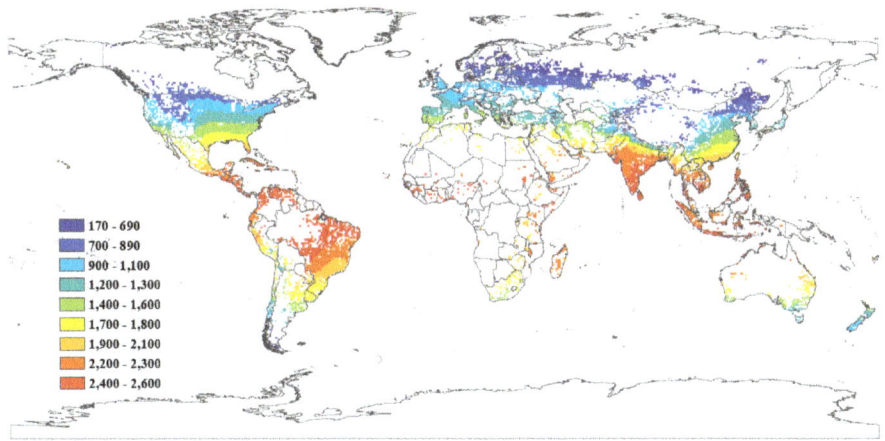

Figure 2. Historical net irrigation water requirement in 1999. The same irrigated area than in 1960 is used the same in Figure 1.

3.2. Net Irrigations Requirements for Future Climate Projections

3.2.1. Climate Data from the CSIRO Model under RCP 4.5 Emissions Scenarios

Figure 3 shows the distribution of IRnet, according to RCP 4.5. As a result, the regions around the equator have *IRnet* between 2400 and 3100 mm/year, while at the level of the Arctic Circle, it is 210 and 990 mm/year. However, North America and the Middle East and the North Africa region have an increase of 15.62% and 11.07% respectively (Table 2). Furthermore, it should be noted that the region of Latin America and the Caribbean has the highest total of IRnet with more than 1.48 km^3/year, followed by the region of East Asia and the Pacific and the region of North America.

Table 2. CSIRO regional net irrigation water.

RCP	Region	Surface (km^2)	Min (mm/yr)	Max (mm/yr)	Mean (mm/yr)	Change of Irnet (%)	Total of Irnet (km^3/yr)
4.5	Latin America and Caribbean	673.68	759.35	3120.93	2199.48	10.38	1.48
	South Asia	327.56	720.48	2597.77	2034.26	10.80	0.67
	Sub-Saharan Africa	63.39	1515.65	2610.28	2161.83	6.57	0.14
	Europe and Central Asia	395.01	239.24	2572.42	940.88	7.69	0.37
	Middle East and North Africa	154.29	1089.54	2500.36	1792.10	11.07	0.28
	East Asia and Pacific	709.39	482.16	2518.66	1547.89	10.18	1.10
	North America	729.38	211.97	2220.06	1278.71	15.62	0.93
8.5	Latin America and Caribbean	673.68	751.64	3030.97	2175.67	9.19	1.47
	South Asia	327.56	772.23	2597.77	1999.00	8.88	0.65
	Sub-Saharan Africa	63.39	1539.50	2702.80	2190.63	7.99	0.14
	Europe and Central Asia	395.01	247.37	1814.57	976.65	11.79	0.39
	Middle East and North Africa	154.29	1149.25	2409.28	1747.95	8.33	0.27
	East Asia and Pacific	709.39	512.35	2529.23	1528.12	8.78	1.08
	North America	729.38	249.81	2226.86	1290.68	16.71	0.94

Same than Table 1.

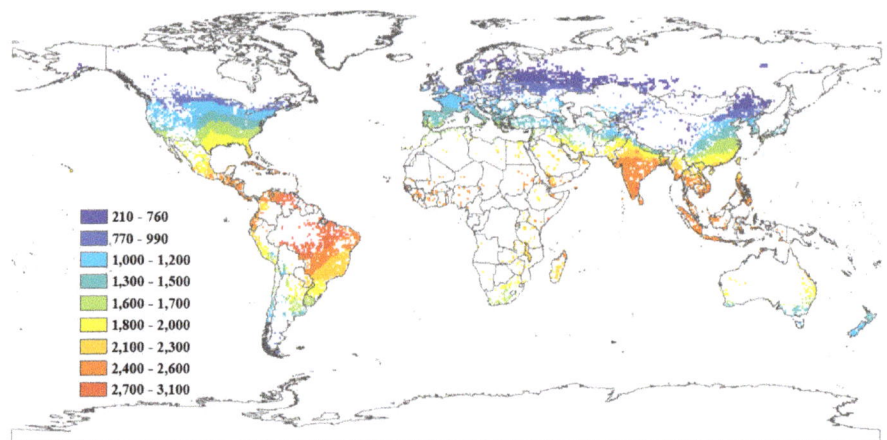

Figure 3. Net irrigation water requirement of CSIRO model in 2050 (RCP4.5). The same irrigated area than in 1960 is used the same in Figure 1.

3.2.2. Climate Data from the CSIRO Model under RCP 8.5 Emissions Scenarios

According to RCP 8.5, Latin America and the Caribbean have the highest IRnet quantity at 1.47 km^3/year, followed by the East and Pacific region and the region of North America. However, the North American region is expected to have the most significant increase with more than 16% (Table 2) compared to 1960, followed by Europe and Central Asia (11.79%). Figure 4 shows the distribution of IRnet. As a result, the countries around the equator have an Irnet between 2300 and 3000 mm/year, while around the Arctic Circle, it is 250 to 990 mm/year.

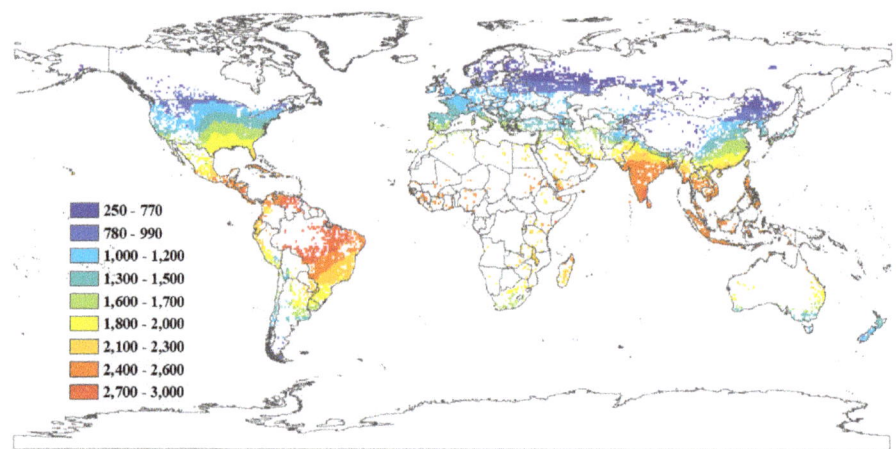

Figure 4. Net irrigation water requirement of CSIRO model in 2050 (RCP8.5). The same irrigated area than in 1960 is used the same in Figure 1.

3.2.3. Climate Data from the ECHAM Model under RCP 4.5 Emissions Scenarios

In Figure 5, the value of the IRnet is between 240 and 2800 mm/year. The countries above the latitude of Cancer have an Irnet between 240 and 2000 mm/year. According to RCP 4.5, the Latin American and Caribbean region has a large Irnet total of 1.43 km^3/year (Table 3), followed by the North American region, and the region of Europe and Central Asia. However, the region of Europe and Central Asia increased by more than 16% compared with 1960, followed by the North American region, with an increase of 11.83%.

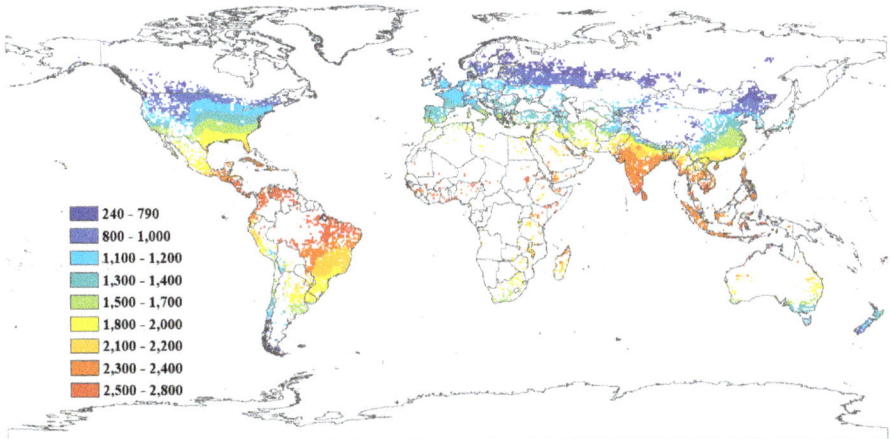

Figure 5. Net irrigation water requirement of ECHAM model in 2050 (RCP4.5). The same irrigated area than 1960 is used the same in Figure 1.

3.2.4. Climate Data from the ECHAM Model under RCP 8.5 Emissions Scenarios

Figure 6 shows that the amount of IRnet varies between 260 and 2800 mm/year. The small quantities are mostly observed around the poles. However, between Cancer and Capricorn latitude, the values are quite high between 2300 and 2800 mm/yr. Compared with 1960, the region of the Middle East and North Africa and part of Europe and Central Asia have an increase of more than 13%, followed by North America (10.93%). However, the region of Latin America and the Caribbean has the highest total of IRnet with a quantity of more than 1.45 km^3/year, followed by the area of East Asia and the Pacific with over 1.12 km^3/year (Table 3).

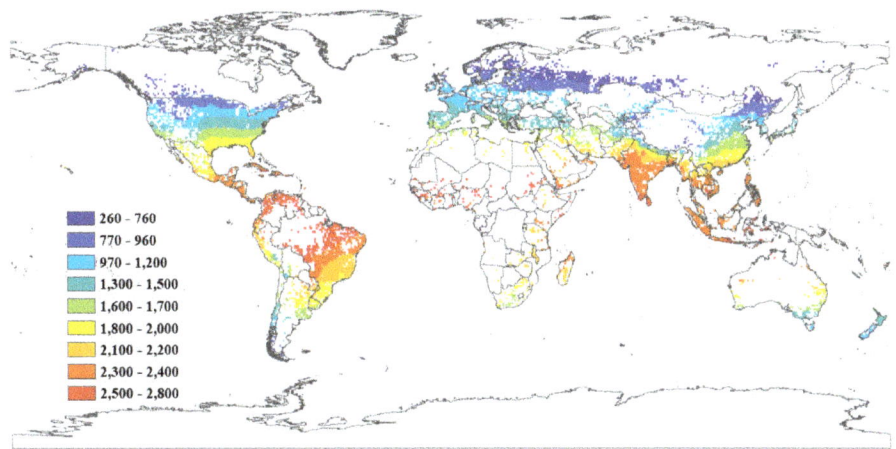

Figure 6. Net irrigation water requirement of ECHAM model in 2050 (RCP8.5). The same irrigated area than in 1960 is used the same in Figure 1.

Table 3. ECHAM regional net irrigation water requirement.

RCP	Region	Surface (km²)	Min (mm/yr)	Max (mm/yr)	Mean (mm/yr)	Change of Irnet (%)	Total of Irnet (km³/yr)
4.5	Latin America and Caribbean	673.68	702.98	2828.20	2116.19	6.20	1.43
	South Asia	327.56	706.12	2537.35	1997.27	8.78	0.65
	Sub-Saharan Africa	63.39	1501.70	2679.44	2187.80	7.85	0.14
	Europe and Central Asia	395.01	242.69	1725.62	1015.36	16.22	0.40
	Middle East and North Africa	154.29	1151.41	2407.62	1772.93	9.88	0.27
	East Asia and Pacific	709.39	470.50	2519.95	1522.59	8.38	1.08
	North America	729.38	328.15	2226.51	1236.71	11.83	0.90
8.5	Latin America and Caribbean	673.68	723.29	2808.93	2151.61	7.98	1.45
	South Asia	327.56	738.32	2538.45	2024.69	10.28	0.66
	Sub-Saharan Africa	63.39	1563.68	2683.62	2212.08	9.04	0.14
	Europe and Central Asia	395.01	262.36	1796.76	988.77	13.17	0.39
	Middle East and North Africa	154.29	1181.75	2451.36	1825.45	13.13	0.28
	East Asia and Pacific	709.39	457.14	2534.39	1577.15	12.27	1.12
	North America	729.38	287.19	2252.33	1226.80	10.93	0.89

Same than Table 1.

3.2.5. Climate Data from the GFDL Model under RCP 4.5 Emissions Scenarios

According to RCP 4.5, the quantity of IRnet varies between 230 and 2600 mm/year (Figure 7), and the countries located near the equator have the most considerable amounts between 2200 and 2600 mm/year. However, those situated, for example, above the latitude of Cancer have relatively low values, notably the countries near the arctic polar circle with quantities between 230 and 890 mm/year. However, there is an increase of approximately 7.37% (Table 4) in the European and Central Asian region. For example, the North American region is expected to increase by approximately 0.75%. However, the Latin American and Caribbean region has the highest IRnet total at 1.39 km³/year, while sub-Saharan Africa has a total of 0.13 km³/year.

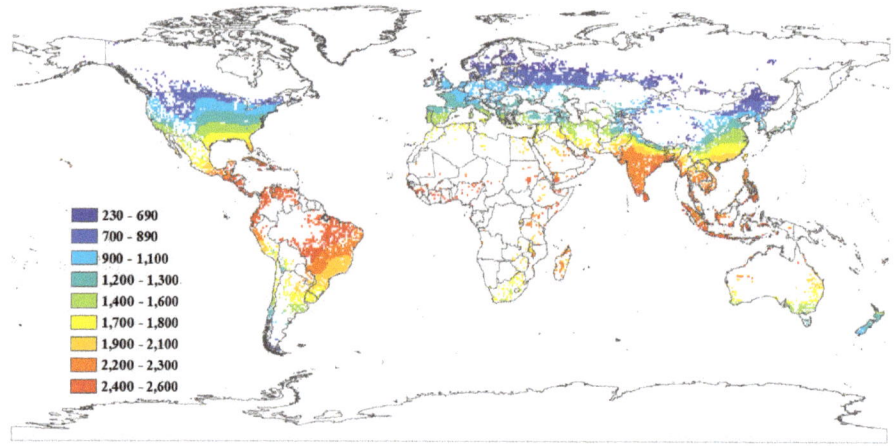

Figure 7. Net irrigation water requirement of GFDL model in 2050 (RCP4.5). The same irrigated area than in 1960 is used the same in Figure 1.

Table 4. GFDL regional net irrigation water requirement.

RCP	Region	Surface (km²)	Min (mm/yr)	Max (mm/yr)	Mean (mm/yr)	Change of Irnet (%)	Total of Irnet (km³/yr)
4.5	Latin America and Caribbean	673.68	766.57	2610.37	2070.56	3.91	1.39
	South Asia	327.56	722.69	2396.12	1898.11	3.38	0.62
	Sub-Saharan Africa	63.39	1421.08	2503.05	2103.97	3.71	0.13
	Europe and Central Asia	395.01	233.90	1852.88	938.02	7.37	0.37
	Middle East and North Africa	154.29	1152.52	2385.55	1724.59	6.88	0.27
	East Asia and Pacific	709.39	431.42	2457.20	1453.15	3.44	1.03
	North America	729.38	375.57	2118.83	1114.22	0.75	0.81
8.5	Latin America and Caribbean	673.68	740.60	2728.36	2121.70	6.48	1.43
	South Asia	327.56	478.13	2455.81	1975.89	7.62	0.65
	Sub-Saharan Africa	63.39	1483.76	2546.20	2110.88	4.06	0.13
	Europe and Central Asia	395.01	251.24	1819.90	972.12	11.27	0.38
	Middle East and North Africa	154.29	1073.82	2430.30	1709.14	5.93	0.26
	East Asia and Pacific	709.39	445.78	2834.22	1491.47	6.17	1.06
	North America	729.38	402.83	2224.01	1172.29	6.00	0.86

Same than Table 1.

3.2.6. Climate Data from the GFDL Model under RCP 8.5 Emissions Scenarios

For RCP 8.5, the amount of IRnet varies from 250 to 2700 mm/year (Figure 8). However, the region of East Asia and the Pacific has a total of 1.06 km³/year behind the Latin American and Caribbean region (1.13 km³/year), while the region of sub-Saharan Africa has a total of 0.13 Km³/year (Table 4). However, the European region has an increase of 11.27% over 1960 followed by South Asia (7.62%), the region of sub-Saharan Africa, for example, has a rise of 4.06%. The countries around the equator have a relatively large quantity of IRnet. These quantities vary from 2100 to 2700 mm/year. On the other hand, and the countries around the poles have *IRnet* between 310 and 910 mm/year.

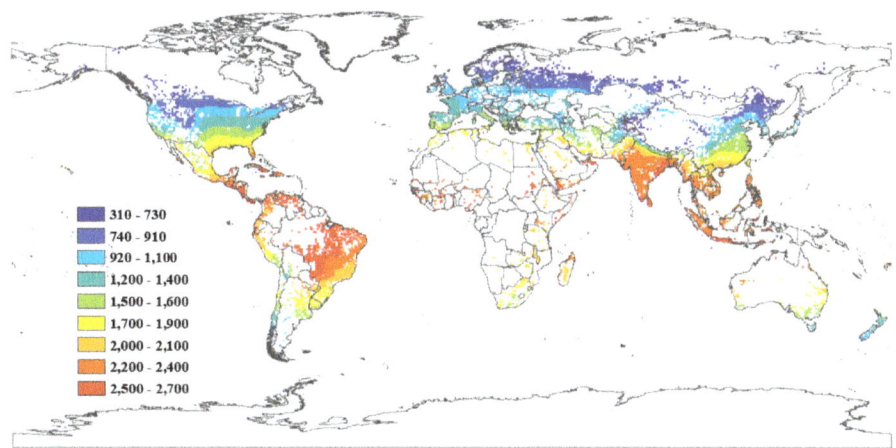

Figure 8. Net irrigation water requirement of GFDL model in 2050 (RCP8.5). The same irrigated area than in 1960 is used the same in Figure 1.

3.2.7. Climate Data from the MIROC 5 Model under RCP 4.5 Emissions Scenarios

According to RCP 4.5, the range of IRnet varies between 310 and 2700 mm/year. The countries situated between Cancer and Capricorn have the highest quantities, between 2300 and 2700 mm/year (Figure 9). On the other hand, countries outside these latitudes and near poles have an amount between 310 and 1000 mm/year. The region of Latin America and the Caribbean has a total IRnet of 1.44 km³/year (Table 5), followed by the East Asian and Pacific region and the North American region.

It should be added, however, that the North American part shows an increase of more than 24% while the sub-Saharan Africa region is likely to increase by only 8.37%.

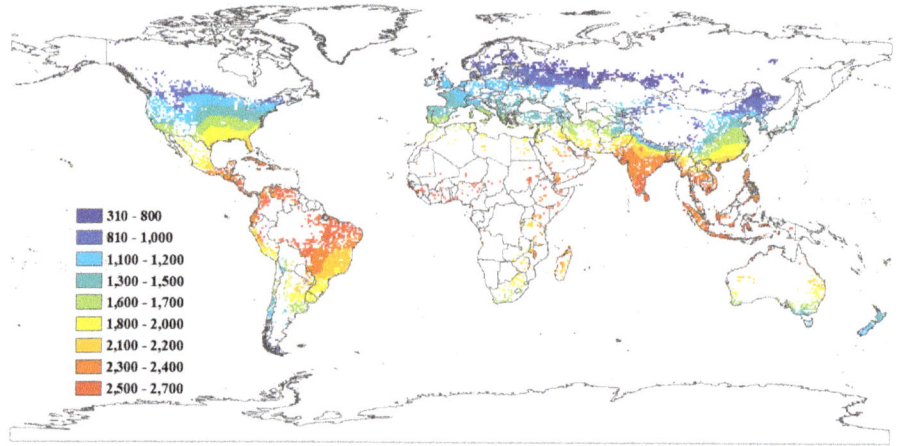

Figure 9. Net irrigation water requirement of MIROC5 model in 2050 (RCP4.5). The same irrigated area than in 1960 is used the same in Figure 1.

Table 5. MIROC 5 regional net irrigation water requirement.

RCP	Region	Surface (km²)	Min (mm/yr)	Max (mm/yr)	Mean (mm/yr)	Change of Irnet (%)	Total of Irnet (km³/yr)
4.5	Latin America and Caribbean	673.68	808.48	2723.54	2139.77	7.39	1.44
	South Asia	327.56	629.89	2481.52	2003.80	9.16	0.66
	Sub-Saharan Africa	63.39	1549.44	2699.46	2215.85	8.70	0.14
	Europe and Central Asia	395.01	306.89	1801.47	1035.56	18.53	0.41
	Middle East and North Africa	154.29	1214.59	2538.07	1852.36	14.81	0.29
	East Asia and Pacific	709.39	510.65	2506.92	1565.36	11.43	1.11
	North America	729.38	309.59	2207.78	1337.30	20.92	0.98
8.5	Latin America and Caribbean	673.68	805.32	2682.33	2135.46	7.17	1.44
	South Asia	327.56	644.63	2477.88	2016.45	9.85	0.66
	Sub-Saharan Africa	63.39	1585.26	2703.02	2209.22	8.37	0.14
	Europe and Central Asia	395.01	334.81	1808.63	1050.90	20.29	0.42
	Middle East and North Africa	154.29	1268.99	2515.18	1882.25	16.65	0.29
	East Asia and Pacific	709.39	548.55	2519.20	1574.44	12.07	1.12
	North America	729.38	300.50	2211.16	1375.57	24.38	1.00

Same than Table 1.

3.2.8. Climate Data from the MIROC 5 Model under RCP 8.5 Emissions Scenarios

With a total of 1.44 km³/year, the region of Latin America and the Caribbean has the most significant amount of IRnet (Table 5), followed by part of East Asia and the Pacific (1.12 km³/year) and the area of North America (1 km³/year). The countries close to the equator have the most substantial quantities between 2200 and 2700 mm/year (Figure 10). It should be added, however, that the North American region shows an increase of 24.38% followed by the region of Europe and Central Asia with an increase of 20.29%.

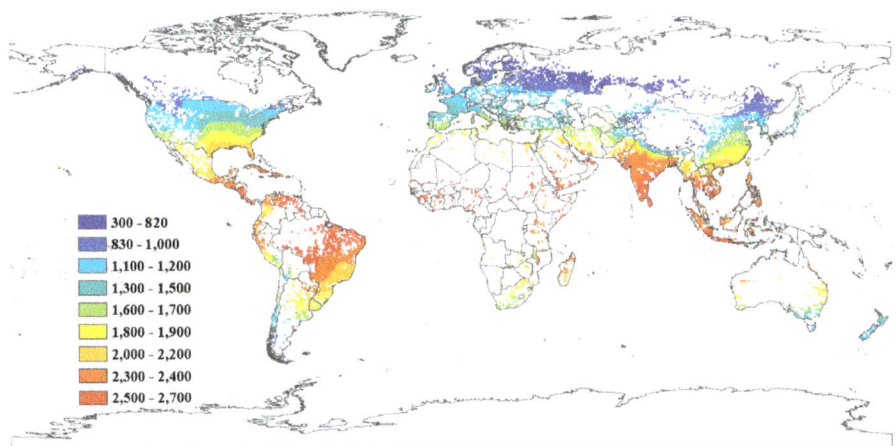

Figure 10. Net irrigation water requirement of MIROC5 model in 2050 (RCP8.5). The same irrigated area than in 1960 is used the same in Figure 1.

3.2.9. Climate Data from the NCAR Model under RCP 4.5 Emissions Scenarios

According to RCP 4.5, the region of Latin America and Caribbean has a total net share of 1,699,770 mm/year, followed by the East Asian and Pacific region (Table 6). The countries near the polar circles have an IRnet between 0 and 270 mm/year, while those between Cancer and Capricorn have an IRnet between 1300 and 1700 mm/year (Figure 11). RCP 4.5 shows a decrease of more than 57% in the region of Europe and Central Asia, more than 46% in the North American area, and over 39% in the region of sub-Saharan Africa.

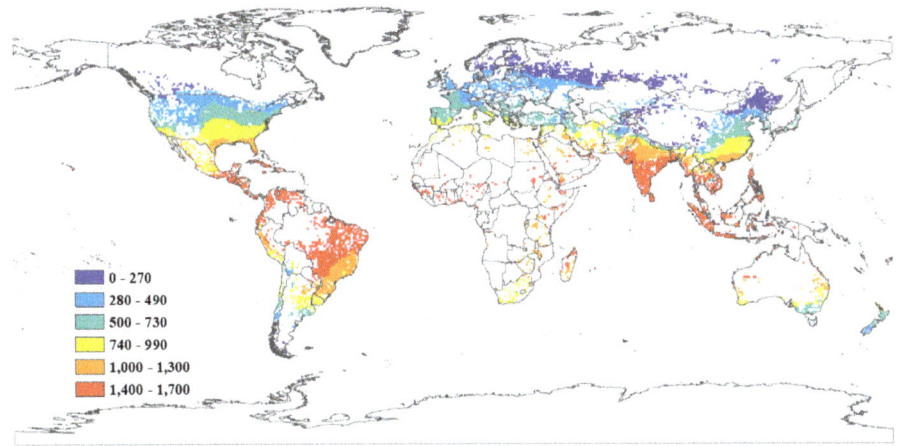

Figure 11. Net irrigation water requirement of NCAR model in 2050 (RCP4.5). The same irrigated area than in 1960 is used the same in Figure 1.

3.2.10. Climate Data from the NCAR Model under RCP 8.5 Emissions Scenarios

Figure 12 shows the variation of IRnet according to RCP 8.5 with extremes between 2200 and 2700 mm/year around the equator and the minimums between 170 and 1000 mm/year near the polar circles. However, it is the region of Latin America and the Caribbean which has the most significant amount of IRnet with more than 1.43 km^3/year, followed by part of North America and the region of

East Asia and the Pacific. Besides, the area of North America shows an increase of over 68% (Table 6), followed by Europe and Central Asia region with 23.20%, while the area of sub-Saharan Africa has only 5.66%.

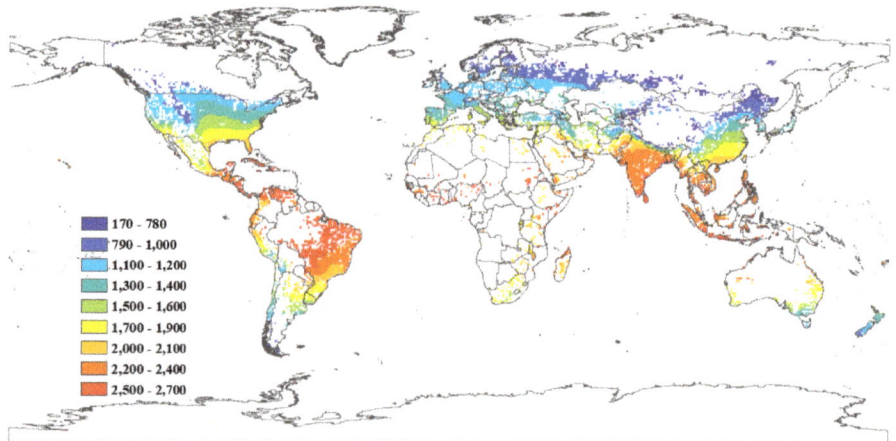

Figure 12. Net irrigation water requirement of NCAR model in 2050 (RCP8.5). The same irrigated area than in 1960 is used the same in Figure 1.

Table 6. NCAR regional net irrigation water requirement.

RCP	Region	Surface (Km^2)	Min (mm/yr)	Max (mm/yr)	Mean (mm/yr)	Change of Irnet (%)	Total of Irnet (km^3/yr)
4.5	Latin America and Caribbean	673.68	170.73	1677.69	1201.25	-39.72	0.81
	South Asia	327.56	0.00	1521.43	1085.52	-40.88	0.36
	Sub-Saharan Africa	63.39	707.83	1601.78	1228.65	-39.43	0.08
	Europe and Central Asia	395.01	0.00	1016.78	368.18	-57.86	0.15
	Middle East and North Africa	154.29	406.59	1458.61	936.02	-41.99	0.14
	East Asia and Pacific	709.39	0.00	1508.02	732.07	-47.89	0.52
	North America	729.38	0.00	1325.51	591.91	-46.48	0.43
8.5	Latin America and Caribbean	673.68	716.76	2691.60	2122.18	6.50	1.43
	South Asia	327.56	380.50	2523.66	2002.29	9.06	0.66
	Sub-Saharan Africa	63.39	1520.56	2643.69	2143.38	5.66	0.14
	Europe and Central Asia	395.01	168.03	1783.16	1076.38	23.20	0.43
	Middle East and North Africa	154.29	1090.51	2350.35	1746.61	8.25	0.27
	East Asia and Pacific	709.39	421.13	2563.41	1527.91	8.76	1.08
	North America	729.38	381.65	2247.39	1865.74	68.70	1.36

Same than Table 1.

4. Discussion

This global study examines and quantifies the impacts of climate change on the net irrigation needs of IRnet maize and thus provides new information required for an optional production of maize. The harshness of the effect of climate change on IRnet is evaluated by comparing the impact of climate change on IRnet in the year 1960 and the year 2050.

By examining the results of the process of this paper (Figure 13), it appears that between 1960 and 1999, IRnet's overall average was 1550.74 mm/year in 1960, compared to 1587.55 mm/year in 1999. This represents an increase in the global IRnet average of approximately 2%. The same observation can be made by comparing the means of the global IRnet in 1960 with the other five models. For example, for CSIRO, RCP 4.5 presents a total average of 1707.88 mm/year, which is an overall increase of just over 10%.

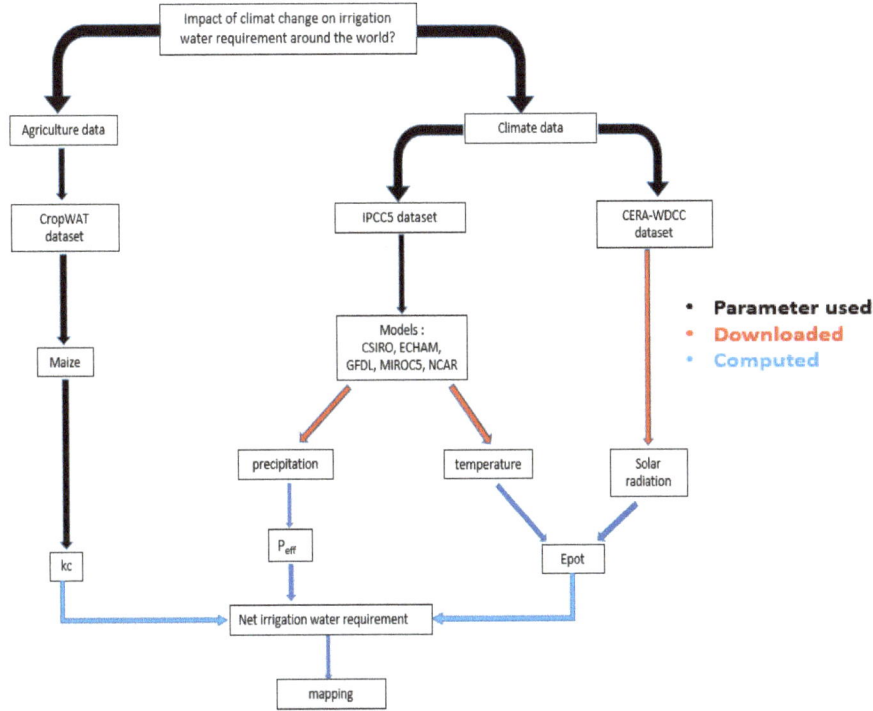

Figure 13. Work chart. Diagram of our work which briefly shows the process of our work.

However, according to most of the five GCMs, the sub-Saharan Africa region is the region with the highest average of Irnet. The values range from 2110.88 mm/year (GFDL) to 2209 mm/year (MIROC 5) according to RCP 8.5, while RC 4.5 has values ranging from 1228.65 mm/year (NCAR) to 2215 mm/year (ECHAM). However, the sub-Saharan Africa region is also one of the regions with the smallest variation of Irnet. Indeed, this region is characterized by its low rainfall and also often confronted with drought. Most of the countries in this area are generally developing countries, which explains the low expansion of irrigation. Nevertheless, maize production covers a vast area in Nigeria. It is 7th in the world, followed by Tanzania and South Africa. The highest producers are South Africa, which is 9th in the world, followed by Nigeria, and Ethiopia [49]. According to FAO in 1995, Nigeria produced 6.93 million tonnes of maize (Mt), followed by South Africa with 4.87 Mt and Ethiopia with 1.99 Mt. For parts that were equipped for irrigation in 1995, to maintain efficient production according to the result in 2050, Nigeria should reach an IRnet of 2688 mm/year (CSIRO), against 2333.17 mm/year (CSIRO) for Ethiopia and 1858.15 mm/year (CSIRO) for South Africa. The countries such as Niger, Mali, or even Chad are at the gateway to the desert (Sahara), and they produced respectively 2000 t, 266,136 t and 62,537 t in 1995. Those countries are expected to improve their production in 2050, by reaching an IRnet of approximately 2662.14 mm/year for Niger (CSIRO), 2773.05 mm/year for Mali (CSIRO) and 2754.60 mm/year for Tchad (CSIRO). It can also be the cause of the unfortunate practice of irrigation. The sub-Saharan African region has the smallest area of irrigation according to the FAO. As a result, the total water quantity used for irrigation in this region is also the lowest, as predicted by the different models.

However, the Middle East and North African region has a much lower average than the sub-Saharan African region. According to RCP 4.5, the region has an average of approximately 1615.6 mm/year, while the RCP 8.5 predicts an average of 1782.36 mm/year. The North African and Middle East region have better rainfall conditions than the sub-Saharan region. As a result, the amount of water associated with irrigation is much higher. In sub-Saharan Africa, maize is the foremost widely grown crop and could be stapled nourishment for an assessed 50% of the populace [49]. However, the area of the irrigation areas is much larger than that of the sub-Saharan African region, so maize cultivation in the Middle East and North Africa region requires more significant amounts of water due to the increase of the irrigation area. It should also be added that this region has a much more advanced economy than that of the sub-Saharan African region. In 1995, maize production in this part of the world, according to FAO, was negated by Egypt (4.5 Mt), followed by Iran (697,246 t) and Syria (199,000 t). In 2050, according to ECHAM RPC 8.5 for example, to increase the production in this region, the amount of IRnet should be approximately 2127.62 mm/year in Egypt, followed by Syria with 1829.01 mm/year and 1828.41 mm/year in Iran. However, it should be noted that these different quantities concern only the areas equipped for irrigation in 1995.

The region of Europe and Central Asia is one of the areas with the lowest average of Irnet. In spite of a reasonably large surface area, this region uses a minimal amount of water for irrigation. The proportion varies from 859.57 mm/year for RCP 4.5 to 1012.92 mm/year for RCP 8.5. This region is most often plentiful in rainfall, and in most countries, it rains throughout the year in general. It should similarly be pointed out that most of this region comprises developed countries, hence the extension of the irrigation areas in this region. Unlike countries in the sub-Saharan African region, most countries do not experience more than three or four months of the rainy season. It shows, on the one hand, the difference in the irrigation surface, but also, on the other hand, the quantity of water used for irrigation. According to Eurostat, France and Romania alone accounted for approximately 45% of maize production in Europe in 2017. However, in 1995, according to FAO, France's production was 12.58 Mt and 9.92 Mt for Romania.

On the other hand, in central Asia, according to the FAO, Afghanistan remains the leading producer of maize in this zone. In 1995, maize production in Afghanistan was 530,000 t. To maintain this production or in the perspective of increasing it, France should reach an average IRnet of 1254.55 mm/year in 2050, according to MIROC 5 RPC 8.5. However, according to the same GCM, Romania should have an IRnet of 1308.17 mm/year against 1462.08 mm/year for Afghanistan, in 2050.

The region of South Asia has a smaller irrigated area than the region of Europe and Central Asia. However, the South Asian region uses a large amount of water for irrigation. The precipitations in this region are not as abundant as that in the European and Central Asian region. Nevertheless, the countries of the South Asian region have economies that are solid enough to support the costs of irrigation, but also the existence of several water sources to meet the water needs of irrigation. Unlike the region of sub-Saharan Africa, water sources for irrigation are not always accessible or nonexistent. According to the models, South Asia is one of the regions that is expected to be moderately affected by the effects of climate change. Most of the models estimate that the amount of water used for maize growing will be between 0.6 and 0.62 km^3/year. This total amount of Irnet is more significant than that of the region of Europe and Central Asia. Maize production in this region is dominated by India with the production of 9.53 Mt in 1995 against 12.04 Mt in 2000 according to the FAO. In 2050, MIROC 5 RCP 8.5 predicts an IRnet of approximately 2151.82 mm/year. Pakistan was the second producer with a production of 1.50 Mt in 1995 and 1.65 Mt in 2000. The prediction of MIROC 5 is 1879.68 mm/year, to ensure a better production in 2050.

The region of Latin America and the Caribbean is the region that uses the most considerable amount of water for irrigation, according to almost all models. The total irrigation values vary from 0.81 km^3/year to 1.48 km^3/year. The total area of irrigation is expected to be approximately 673.68 km^2 far in front of the sub-region of Africa or the region of Europe, and in Central Asia or even in the South Asian area. It should be pointed out that the region of Latin America and the Caribbean is often confronted with drought, yet this region is full of enough water source but also a more or less stable economy to meet the demands of irrigation. However, the Latin American and Caribbean region uses large quantities of water for irrigation because it has countries with the maximum IRnet used worldwide, with values of up to 3030 mm/year according to GCS CSIRO RCP 8.5. Ranking third and eighth as world maize producers in 2019 by the Index Mundin, Brazil and Mexico are the top two maize producers in Latin America and Caribbean. in 1995. The production of Brazil was 36.27 Mt, against 18.35 Mt for Mexico. To ensure sustainable production, the GCM ECHAM RCP 8.5 predicts an average Irnet of 2465.46 mm/year for 2050 and 2171.20 mm/year for Mexico.

East Asia and Pacific is the second region with the largest area irrigated, just behind the North American region. China holds most of this area. The East Asian and Pacific region also ranks second for the total amount of water used in irrigation (Irnet), behind the Latin American and Caribbean region. However, this region has substantial quantities of the water source. It enables it to cope with the high demand for irrigation water in most of the countries of this region. It must also be added that the countries of this region have, for the most part, an economy powerful enough to meet the requirements of their irrigation. There is notably China, which is the first economic power in the world. According to MO Xing-Guo [50], the potential evapotranspiration is expected to increase by 8%–16% and 7%–10% in the 2050s. However, the 2019 Index Mundin rankings reported that China and Indonesia are respectively the second and the eleventh world maize producers. As a result, they are the first producers of this region. According to the FAO, maize production in 1995 was 112.36 Mt and 8.25 Mt for Indonesia. Given the growth in demography, especially in China, and to ensure better production, China is expected to, according to ECHAM RPC 8.5, reach an Irnet of 1433.3 mm/year in 2050. For Indonesia, in 2050, the Irnet is expected to be 2534.72 mm/year. However, the East Asian and Pacific region is expected to experience an increase in Irnet quantity from 1960.

The North American region is the region with the largest irrigated area. This area is approximately 729 km^2, placing this region first in the East Asian and Pacific region. As this large area floods, the North American region uses a total Irnet amount of 0.81 km^3/year at 1 km^3/year. This region also has one of the smallest Irnet averages. The North American region is a region with good rainfall. This region is one of the areas that has, more or less, water throughout the year. It is also necessary to add the existence of important water resources. As a result, despite the vast area of irrigated areas, the region of North America, because of the pluviometry abundance it possesses, can cope with the water requirements of irrigation. As the world's largest producer, the United States of America dominates maize production in the North American region. Its production amounts to more than 380 Mt in 2019. According to the FAO, the production of maize was 187.97 mm/year in 2000 and the production reached 251.85 Mt. According to the NCAR RCP 8.5, the United States of America is expected to foresee an Irnet of 1503.05 mm/year in 2050 to ensure production. It should be added, however, that the economy of this region is one of the most powerful in the world, which could explain the expansion of the irrigation areas.

According to RCP 4.5, these increases vary between 0.74% (North America) and 20.92% (North America), while the RCP 8.5 predicts increases of 4.06% (sub-Saharan Africa) to more than 68% (North America). The region of sub-Saharan Africa is expected to have a slight increase according to the different models, perhaps due to its already extreme climate, while North America is the region with the most variation due to climate change. However, NCAR's RCP 4.5 looks somewhat like an overall decrease of *IRnet*. This decrease varies from more than 39% (sub-Saharan Africa) to more than 57% (Europe and Central Asia).

To consider the value of the importance of this study, it is essential to talk about the many sources of uncertainty. In general, GCMs do not take into account the influence of CO_2 on plant physiology. As a result, there is an underestimation of regional warming and an overestimation of humidity, particularly in the tropics. Further, the modeling of the wetlands in climate change research on water wants is not ideal. In this study, only one plant was used for the simulation. However, if the simulation was carried out with a multicultural mode, this would indicate that these areas are more favored by climate change than other irrigated areas in the same region of the world. However, determining the impact of climate change on the water requirements of irrigation requires the inclusion of the different indirect effects of climate change. Moreover, the economic conditions are an essential factor in any adaptation, a state that in this study that is not taken into account.

However, the same conclusions have been observed in other works. As Wenfeng Liu [51] has shown in his work, the regions as arid desert hot and cold, arid steppe hot and cold, temperate dry and hot summer, temperate dry and warm summer have a high level of irrigation water requirements. The same results can be observed in the different pictures in our study. Those regions have the most elevated amount of irrigation water requirements. The increase is highest in the RCP 8.5 scenario than RCP 4.5 scenario as Hanqing Xu [52] concludes. If the country is specified, Tianwa Zhou [53] shows that the irrigation water requirement in his study area which is western Inner Mongolia in China is expected to be 648 mm under scenario A2 and 639 mm under B2. These values are included in the predicted interval for this area in 2050 in our study.

5. Conclusions

The analysis of the results shows an increase across the different regions. Fisher [21] estimates that the impact of climate change on IRnet is very significant, with a 20% increase between 2000 and 2080. Based on IPCC-5 data, this study also shows an increase in IRnet between 1960 and 2050. However, the continent of America (the region of North America and the region of Latin America and the Caribbean) is the continent with the most significant increase of *IRnet*. According to the RCP 4.5, the continent of America has the highest total of IRnet which is between 1.24 km^3/yr (NCAR) and 2.42 km^3/yr (MIROC 5). For the RCP 8.5, it is between 2.29 km^3/yr (GFDL) and 2.79 km^3/yr (NCAR). The African continent (the region of sub-Saharan Africa and the Middle East and North Africa) is the continent with the smallest increase in *IRnet* as well as the smallest irrigation area. In this part of the world, the amount of the IRnet is between 0.22 km^3/yr (NCAR) and 0.43 km^3/yr (MIROC) according to RCP 4.5. For RCP 8.5, the amount of the IRnet in Africa is between 0.39 km^3/yr (GFDL) and 0.43 km^3/yr (MIROC 5). On the other hand, the region of North America is the region with the largest area of irrigated area, followed by the region of East Asia and the Pacific and the region of Latin America and the Caribbean. The European and Central Asian region has the lowest average of *IRnet*, followed by the region of East Asia and the Pacific. In other words, the continent of Europe and of Asia are the continents that use the least water globally in irrigation. It should be remembered, that RCP 4.5 of the GCM NCAR predicts, in contrast to other models, an overall decrease in *IRnet* in all regions of the globe. By using maps from different models, decision-makers can indeed observe sensitive areas and thus develop a water distribution policy for more efficient irrigation. Although they show different values, the cards all have, for the most part, one thing in common: They predict an increase in IRnet. Based on the data of the IPCC-5, the different models reveal the zones that will face important increases. This way, decision-makers have to create new water policy strategies to deal with different changes. The previous work is mainly based on the data of the IPCC-4, while this study can be considered as another version of the update of the research carried out on the global representation of the IRnet. Thereby, this study helps decision-makers to make corrections in making their decision, and by interpolation on the other cultivated plants, to affirm that the result will be more or less the same. Specifically, that is to say, that the *IRnet* of the greatest majority of plants of culture will know an important increase, more or less, depending on the region.

Author Contributions: Conceptualization, H.L.; methodology, A.O.A.; software, A.O.A.; validation, A.O.A. and H.L.; formal analysis, A.O.A., Y.A.H. and M.S.; investigation, A.O.A., and Y.Z.; resources, H.L.; data curation, A.O.A.; writing—original draft preparation, A.O.A. and Y.A.H.; writing—review and editing, A.O.A., Y.A.H., M.S., H.L., and Y.Z.; visualization, H.L.; supervision, H.L.; project administration, H.L.; funding acquisition, H.L.

Funding: This research is supported by National Key Research and Development Program (grant Nos. 2018YFA0605400, 2016YFA0601504); NNSF (grant Nos. 41830752 and 41571015); and the open funding of the Laboratory (grant Nos. OFSLRSS201806, HRM201704, 2018B44114).

Acknowledgments: The authors thank the reviewers and editors for their valuable comments about the manuscript.

Conflicts of Interest: The authors declare no conflict of interest.

References

1. Kotir, J.H. Climate change and variability in Sub-Saharan Africa: A review of current and future trends and impacts on agriculture and food security. *Environ. Dev. Sustain.* **2011**, *13*, 587–605. [CrossRef]
2. Rossati, A. Global warming and its health impact. *Int. J. Occup. Env. Med. (IJOEM)* **2017**, 8. [CrossRef] [PubMed]
3. Ebi, K.L.; Ogden, N.H.; Semenza, J.C.; Woodward, A. Detecting and attributing health burdens to climate change. *Environ. Health Perspect.* **2017**, *125*, 085004. [CrossRef] [PubMed]
4. Pecl, G.T.; Araújo, M.B.; Bell, J.D.; Blanchard, J.; Bonebrake, T.C.; Chen, I.-C.; Clark, T.D.; Colwell, R.K.; Danielsen, F.; Evengård, B. Biodiversity redistribution under climate change: Impacts on ecosystems and human well-being. *Science* **2017**, *355*, eaai9214. [CrossRef] [PubMed]
5. Myers, S.S.; Smith, M.R.; Guth, S.; Golden, C.D.; Vaitla, B.; Mueller, N.D.; Dangour, A.D.; Huybers, P. Climate change and global food systems: Potential impacts on food security and undernutrition. *Annu. Rev. Public Health* **2017**, *38*, 259–277. [CrossRef]
6. Serdeczny, O.; Adams, S.; Baarsch, F.; Coumou, D.; Robinson, A.; Hare, W.; Schaeffer, M.; Perrette, M.; Reinhardt, J. Climate change impacts in Sub-Saharan Africa: From physical changes to their social repercussions. *Reg. Environ. Chang.* **2017**, *17*, 1585–1600. [CrossRef]
7. Caney, S. Human rights, responsibilities, and climate change. In *Environmental Rights*; Routledge: London, UK, 2017; pp. 117–137.
8. Moazenzadeh, R.; Mohammadi, B. Assessment of bio-inspired metaheuristic optimisation algorithms for estimating soil temperature. *Geoderma* **2019**, *353*, 152–171. [CrossRef]
9. Ahmadzadeh Araji, H.; Wayayok, A.; Massah Bavani, A.; Amiri, E.; Abdullah, A.F.; Daneshian, J.; Teh, C.B.S. Impacts of climate change on soybean production under different treatments of field experiments considering the uncertainty of general circulation models. *Agric. Water Manag.* **2018**, *205*, 63–71. [CrossRef]
10. Aghelpour, P.; Mohammadi, B.; Biazar, S.M. Long-term monthly average temperature forecasting in some climate types of Iran, using the models SARIMA, SVR, and SVR-FA. *Theor. Appl. Climatol.* **2019**. [CrossRef]
11. Nerem, R.S.; Beckley, B.D.; Fasullo, J.T.; Hamlington, B.D.; Masters, D.; Mitchum, G.T. Climate-change–driven accelerated sea-level rise detected in the altimeter era. *Proc. Natl. Acad. Sci. USA* **2018**, *115*, 2022–2025. [CrossRef]
12. Clapp, J.; Newell, P.; Brent, Z.W. The global political economy of climate change, agriculture and food systems. *J. Peasant Stud.* **2018**, *45*, 80–88. [CrossRef]
13. Tol, R.S. The economic impacts of climate change. *Rev. Environ. Econ. Policy* **2018**, *12*, 4–25. [CrossRef]
14. Hsiang, S.; Kopp, R.; Jina, A.; Rising, J.; Delgado, M.; Mohan, S.; Rasmussen, D.; Muir-Wood, R.; Wilson, P.; Oppenheimer, M. Estimating economic damage from climate change in the United States. *Science* **2017**, *356*, 1362–1369. [CrossRef] [PubMed]
15. Zhang, P.; Zhang, J.; Chen, M. Economic impacts of climate change on agriculture: The importance of additional climatic variables other than temperature and precipitation. *J. Environ. Econ. Manag.* **2017**, *83*, 8–31. [CrossRef]
16. Estrada, F.; Botzen, W.W.; Tol, R.S. A global economic assessment of city policies to reduce climate change impacts. *Nat. Clim. Chang.* **2017**, *7*, 403. [CrossRef]
17. Hallegatte, S.; Rozenberg, J. Climate change through a poverty lens. *Nat. Clim. Chang.* **2017**, *7*, 250. [CrossRef]
18. Sonwa, D.J.; Dieye, A.; El Mzouri, E.-H.; Majule, A.; Mugabe, F.T.; Omolo, N.; Wouapi, H.; Obando, J.; Brooks, N. Drivers of climate risk in African agriculture. *Clim. Dev.* **2017**, *9*, 383–398. [CrossRef]
19. Weber, T.; Haensler, A.; Rechid, D.; Pfeifer, S.; Eggert, B.; Jacob, D. Analyzing regional climate change in africa in a 1.5, 2, and 3 C global warming world. *Earth's Future* **2018**, *6*, 643–655. [CrossRef]
20. Murphy, J.M.; Sexton, D.M.H.; Barnett, D.N.; Jones, G.S.; Webb, M.J.; Collins, M.; Stainforth, D.A. Quantification of modelling uncertainties in a large ensemble of climate change simulations. *Nature* **2004**, *430*, 768. [CrossRef]
21. Fischer, G.; Tubiello, F.N.; van Velthuizen, H.; Wiberg, D.A. Climate change impacts on irrigation water requirements: Effects of mitigation, 1990–2080. *Technol. Forecast. Soc. Chang.* **2007**, *74*, 1083–1107. [CrossRef]
22. Lesley, H.; Michael, A.; Stephanie, T. Climate change and Australia. *Wiley Interdiscip. Rev. Clim. Chang.* **2014**, *5*, 175–197. [CrossRef]

23. Kapetanaki, G.; Rosenzweig, C. Impact of climate change on maize yield in central and northern Greece: A simulation study with CERES-Maize. *Mitig. Adapt. Strateg. Glob. Chang.* **1997**, *1*, 251–271. [CrossRef]
24. Konzmann, M.; Gerten, D.; Heinke, J. Climate impacts on global irrigation requirements under 19 GCMs, simulated with a vegetation and hydrology model. *Hydrol. Sci. J.* **2013**, *58*, 88–105. [CrossRef]
25. Döll, P. Impact of Climate Change and Variability on Irrigation Requirements: A Global Perspective. *Clim. Chang.* **2002**, *54*, 269–293. [CrossRef]
26. Al-Ghobari, H.M.; Dewidar, A.Z. Deficit irrigation and irrigation methods as on-farm strategies to maximize crop water productivity in dry areas. *J. Water Clim. Chang.* **2017**, *9*, 399–409. [CrossRef]
27. Lehn, H.; Simon, L.M.; Oertel, M. Climate Change Impacts on the Water Sector. In *Climate Adaptation Santiago*; Krellenberg, K., Hansjürgens, B., Eds.; Springer: Berlin/Heidelberg, Germany, 2014; pp. 59–79. [CrossRef]
28. Lengoasa, J. Climate variability and change: Impacts on water availability. *Irrig. Drain.* **2016**, *65*, 149–156. [CrossRef]
29. Novoa, D.C. Hydro-economic analysis for water resources management in a changing climate. In *Climate Change and the Sustainable Use of Water Resources*; Springer: Berlin, Germay, 2012; pp. 127–141.
30. Babel, M.S.; Agarwal, A.; Shinde, V.R. Climate Change Impacts on Water Resources and Selected Water Use Sectors. In *Climate Change and Water Resources*; CRC Press: Boca Raton, FL, USA, 2014; pp. 126–169.
31. Hamoud, Y.A.; Guo, X.; Wang, Z.; Shaghaleh, H.; Chen, S.; Hassan, A.; Bakour, A. Effects of irrigation regime and soil clay content and their interaction on the biological yield, nitrogen uptake and nitrogen-use efficiency of rice grown in southern China. *Agric. Water Manag.* **2019**, *213*, 934–946. [CrossRef]
32. Frenken, K.; Gillet, V. *Irrigation Water Requirement and Water Withdrawal by Country*; FAO: Rome, Italy, 2012.
33. Rodríguez Díaz, J.A.; Weatherhead, E.K.; Knox, J.W.; Camacho, E. Climate change impacts on irrigation water requirements in the Guadalquivir river basin in Spain. *Reg. Environ. Chang.* **2007**, *7*, 149–159. [CrossRef]
34. Hamoud, Y.A.; Shaghaleh, H.; Sheteiwy, M.; Guo, X.; Elshaikh, N.A.; Khan, N.U.; Oumarou, A.; Rahim, S.F. Impact of alternative wetting and soil drying and soil clay content on the morphological and physiological traits of rice roots and their relationships to yield and nutrient use-efficiency. *Agric. Water Manag.* **2019**, *223*, 105706. [CrossRef]
35. Richard, M.A.; Brian, H.H.; Stephanie, L.; Neil, L. Effects of global climate change on agriculture: An interpretative review. *Clim. Res.* **1998**, *11*, 19–30.
36. Cline, W.R. Global warming and agriculture. *Financ. Dev.* **2008**, *45*, 23.
37. Karl, T.R.; Trenberth, K.E. Modern Global Climate Change. *Science* **2003**, *302*, 1719–1723. [CrossRef] [PubMed]
38. Boonwichai, S.; Shrestha, S.; Babel, M.S.; Weesakul, S.; Datta, A. Climate change impacts on irrigation water requirement, crop water productivity and rice yield in the Songkhram River Basin, Thailand. *J. Clean. Prod.* **2018**, *198*, 1157–1164. [CrossRef]
39. Ali, S.; Liu, Y.; Ishaq, M.; Shah, T.; Ilyas, A.; Din, I. Climate change and its impact on the yield of major food crops: Evidence from Pakistan. *Foods* **2017**, *6*, 39. [CrossRef]
40. Alhaj Hamoud, Y.; Wang, Z.; Guo, X.; Shaghaleh, H.; Sheteiwy, M.; Chen, S.; Qiu, R.; Elbashier, M. Effect of Irrigation Regimes and Soil Texture on the Potassium Utilization Efficiency of Rice. *Agronomy* **2019**, *9*, 100. [CrossRef]
41. Kang, S.; Hao, X.; Du, T.; Tong, L.; Su, X.; Lu, H.; Li, X.; Huo, Z.; Li, S.; Ding, R. Improving agricultural water productivity to ensure food security in China under changing environment: From research to practice. *Agric. Water Manag.* **2017**, *179*, 5–17. [CrossRef]
42. Davis, K.F.; Rulli, M.C.; Seveso, A.; D'Odorico, P. Increased food production and reduced water use through optimized crop distribution. *Nat. Geosci.* **2017**, *10*, 919. [CrossRef]
43. Yue, Q.; Xu, X.; Hillier, J.; Cheng, K.; Pan, G. Mitigating greenhouse gas emissions in agriculture: From farm production to food consumption. *J. Clean. Prod.* **2017**, *149*, 1011–1019. [CrossRef]
44. Jägermeyr, J.; Gerten, D.; Schaphoff, S.; Heinke, J.; Lucht, W.; Rockström, J. Integrated crop water management might sustainably halve the global food gap. *Environ. Res. Lett.* **2016**, *11*, 025002. [CrossRef]
45. Döll, P.; Siebert, S. Global modeling of irrigation water requirements. *Water Resour. Res.* **2002**, *38*, 1–10. [CrossRef]
46. Hamoud, Y.A.; Guo, X.; Wang, Z.; Chen, S.; Rasoul, G. Effects of irrigation water regime, soil clay content and their combination on growth, yield, and water use efficiency of rice grown in South China. *Int. J. Agric. Biol. Eng.* **2018**, *11*, 144–155.

47. Smith, M. *CROPWAT: A Computer Program for Irrigation Planning and Management*; Food and Agriculture Organization of the United Nations: Rome, Italy, 1992.
48. Raziei, T.; Pereira, L.S. Estimation of ETo with Hargreaves–Samani and FAO-PM temperature methods for a wide range of climates in Iran. *Agric. Water Manag.* **2013**, *121*, 1–18. [CrossRef]
49. Badu-Apraku, B.; Fakorede, M.A.B. Maize in Sub-Saharan Africa: Importance and Production Constraints. In *Advances in Genetic Enhancement of Early and Extra-Early Maize for Sub-Saharan Africa*; Springer: Cham, Germany, 2017; pp. 3–10. [CrossRef]
50. Mo, X.-G.; Hu, S.; Lin, Z.-H.; Liu, S.-X.; Xia, J. Impacts of climate change on agricultural water resources and adaptation on the North China Plain. *Adv. Clim. Chang. Res.* **2017**, *8*, 93–98. [CrossRef]
51. Liu, W.; Yang, H.; Folberth, C.; Wang, X.; Luo, Q.; Schulin, R. Global investigation of impacts of PET methods on simulating crop-water relations for maize. *Agric. For. Meteorol.* **2016**, *221*, 164–175. [CrossRef]
52. Xu, H.; Tian, Z.; He, X.; Wang, J.; Sun, L.; Fischer, G.; Fan, D.; Zhong, H.; Wu, W.; Pope, E.; et al. Future increases in irrigation water requirement challenge the water-food nexus in the northeast farming region of China. *Agric. Water Manag.* **2019**, *213*, 594–604. [CrossRef]
53. Zhou, T.; Wu, P.; Sun, S.; Li, X.; Wang, Y.; Luan, X. Impact of future climate change on regional crop water requirement—A case study of Hetao Irrigation District, China. *Water* **2017**, *9*, 429. [CrossRef]

© 2019 by the authors. Licensee MDPI, Basel, Switzerland. This article is an open access article distributed under the terms and conditions of the Creative Commons Attribution (CC BY) license (http://creativecommons.org/licenses/by/4.0/).

Article

Projected Changes in the Frequency of Peak Flows along the Athabasca River: Sensitivity of Results to Statistical Methods of Analysis

Yonas Dibike [1,*], Hyung-Il Eum [2], Paulin Coulibaly [3] and Joshua Hartmann [1]

1. Environment and Climate Change Canada, Watershed Hydrology and Ecology Research Division, University of Victoria, Victoria, BC V8P 5C2, Canada
2. Alberta Environment and Parks (AEP), Environmental Monitoring and Science Division, Calgary, AB T9K 0K4, Canada
3. Civil Engineering Department and the School of Geography and Earth Sciences, McMaster University, Hamilton, ON L8S 4L8, Canada
* Correspondence: yonas.dibike@canada.ca; Tel.: +1-250-363-8910

Received: 8 May 2019; Accepted: 3 July 2019; Published: 4 July 2019

Abstract: Flows originating from alpine dominated cold region watersheds typically experience extended winter low flows followed by spring snowmelt and summer rainfall driven high flows. In a warmer climate, there will be a temperature-induced shift in precipitation from snowfall towards rain along with changes in precipitation intensity and snowmelt timing, resulting in alterations in the frequency and magnitude of peak flow events. This study examines the potential future changes in the frequency and severity of peak flow events in the Athabasca River watershed in Alberta, Canada. The analysis is based on simulated flow data by the variable infiltration capacity (VIC) hydrologic model driven by statistically downscaled climate change scenarios from the latest coupled model inter-comparison project (CMIP5). The hydrological model projections show an overall increase in mean annual streamflow in the watershed and a corresponding shift in the freshet timing to an earlier period. The river flow is projected to experience increases during the winter and spring seasons and decreases during the summer and early fall seasons, with an overall projected increase in peak flow, especially for low frequency events. Both stationary and non-stationary methods of peak flow analysis, performed at multiple points along the Athabasca River, show that projected changes in the 100-year peak flow event for the high emissions scenario by the 2080s range between 4% and 33% depending on the driving climate models and the statistical method of analysis. A closer examination of the results also reveals that the sensitivity of projected changes in peak flows to the statistical method of frequency analysis is relatively small compared to that resulting from inter-climate model variability.

Keywords: Athabasca River; climate projection; hydrologic modelling; peak-flow; return period; stationary analysis; non-stationary analysis

1. Introduction

Climate variability and changes in cold region watersheds are having significant impacts on the different components of the hydrologic-cycle, such as on snow accumulation and melt, soil moisture and runoff affecting local and regional hydrological regimes. Changes in any of these hydrologic processes, including precipitation intensity, snowmelt runoff and antecedent soil moisture, may cause alterations in frequency and intensity of extreme flows [1,2]. While flash floods are usually generated by intense convective rainfalls that occur in summer, snowmelt-driven extreme flows in cold regions environment are more frequent in spring and early summer [3]. Numerous studies also exist that document river ice-jam related floods that can be produced in cold region environments [4,5]. Physical considerations of climatic change in the form of increased temperature and precipitation suggest increased flood risk in various parts of Canada, especially if there is a corresponding increase in precipitation intensity [6,7]. Therefore, in many cases, projected changes in precipitation and temperature and the resulting shift in snowmelt timing are expected to cause changes in the magnitude and timing of peak flow events [8].

Flood frequency analysis has generally been used to model peak flows under the stationary assumption [9]; however, with a changing climate, the assumption of stationarity is being challenged, and a non-stationary flood frequency analysis approaches are becoming more prominent [10,11]. The non-stationarity of the hydro-meteorological series has become important as the water cycle is significantly affected by climate and land use changes (such as deforestation and/or urbanization) and is often characterized by the presence of a trend component (i.e., either linear or non-linear) and/or a sudden jump in the statistical characteristics of data [12]. Cunderlik and Burn [13] emphasized that the presence of significant non-stationarity in a hydrologic time series cannot be ignored when estimating design values for future time horizons. They also showed that ignoring even a weakly significant non-stationarity in the data series may seriously bias the quantile predicted for time horizons as near as 0–20 years in the future. Tan and Gan's [14] investigation of the long-term annual maximum streamflow (AMS) records at 145 stations over Canada also concluded that non-stationary frequency analysis, instead of the traditional stationary approach, should be employed in the future. They have also demonstrated that the non-stationary characteristics of AMS can be accounted by fitting the data to probability distributions with time varying parameters or distribution parameters varying with other factors such as climate anomalies, and land-use change descriptors representing the physical explanations behind various types of non-stationarities found in the streamflow series. However, Ouarda and El-Adlouni [15] have cautioned to use such models with care when the covariate is considered to be time as the direct extrapolation of the currently observed trends can be misleading and lead to erroneous results.

Lopez and Frances [16] have applied two approaches to non-stationary modelling of the annual maximum flood records of 20 continental Spanish rivers. The first approach, where the distribution parameters were modelled as a function of time, only showed the presence of clear non-stationarities in the extreme flow regime; while the second approach, where the parameters are modelled as functions of climate and reservoir indices, highlighted the important role of inter-annual climate variability and reservoir regulation strategies, when modelling the flood regime in continental Spanish rivers. The application of non-stationary analysis in their study also showed that the differences between the non-stationary quantiles and their stationary equivalents might be important over long periods of time and the inclusion of external covariates permits the use of these models as predictive tools. Results of a similar study by Li and Tan [17] that considers the effects of climate variability and reservoir operation in the Daqinghe river basin in China highlighted the necessity of flood frequency analysis under non-stationary conditions, and even suggested possible adoption of alternative definitions of the return period. Seidou et al. [18] have also shown that by using the non-stationary distribution, with a location parameter linked to the maximum nine-day average flow, a much better estimation of flood quantiles is provided than when applying a stationary frequency analysis to the simulated peak flows and flood quantiles (simulated using the non-stationary distribution display the same trends as that of the observed data during the study period). Zhang et al. [19] applied univariate and bivariate models to investigate the nonstationary frequency of flood peak and volume of the Wangkuai Reservoir in China with distribution parameters changing over time. Dong et al. [20] also developed nonstationary bivariate models, where distribution parameters vary with possible physical covariates (i.e., precipitation, urbanization, and deforestation) to model the nonstationary behavior of the flood characteristics of the Dongnai River in Vietnam.

A recent study by Shrestha et al. [21] has presented an assessment of potential impacts of climate change on extreme events in the Fraser River in Canada using model simulated streamflow corresponding to future climate projections. By explicitly considering the non-stationarity of extreme events and quantifying the transient response of peak flow discharge magnitude and frequency to external climate forcing, the study found potential increases in the moderately high (2–20-year return period) streamflow events, while the results were inconclusive for low frequency events (100–200-year return period). Projections from several global and regional climate models over the Athabasca watershed in Canada also show an average change toward more drought-like summer and slightly wetter annual conditions over the region [22,23]. Other studies also agree in a projected decrease in the winter snow accumulation and summer flows, as well as earlier onset of spring freshet in the region [24,25]. Eum et al. [25] reported projected increases in the mean-annual maximum flow at a number of stations along the Athabasca River, although they were statistically significant only at the stations located along the lower reaches. However, those studies have not looked explicitly at projected changes in the frequency and magnitude of peak flow events in the river. Therefore, the main objective of this study is to investigate projected changes in the frequency and severity of peak flow events at various locations along the Athabasca River using multiple stationary and non-stationary flood frequency analysis techniques. This includes exploring the inter-model variability of the results with respect to different climatic-drivers originating from different Global Climate Models (GCMs) and examining its sensitivity to the different statistical methods of flood frequency analysis. This objective is achieved by analyzing the projected changes in the hydrologic regime, and the corresponding peak flows of the Athabasca River as it has been simulated by the Variable Infiltration Capacity (VIC) hydrologic model driven by a select-set of statistically downscaled climate change scenario data derived from the latest coupled model inter-comparison project (CMIP5).

2. Materials and Methods

2.1. Study Area

The Athabasca River basin (ARB, see Figure 1) originates in the Canadian Rockies from the Athabasca Glacier, at over 3700 m above mean sea level (amsl), and flows approximately 1500 km north-eastward through the province of Alberta. It passes by, or through Jasper, Hinton, Whitecourt, Athabasca, and Fort McMurray, before emptying into Lake Athabasca (average elevation ~208 m amsl), which outflows through the Slave River and Lake to the Mackenzie River system. Its total drainage area attains approximately 156,000 km^2 near Old Fort before it flows into Lake Athabasca. The watershed includes various land cover types, such as snow-capped mountains, agricultural plains, boreal forest, wetlands, and small urban areas. The boreal forest is dominated by coniferous followed by mixed and transitional forest. Mean annual precipitation in the watershed ranges from around 300 mm at the downstream end near Lake Athabasca to over 1000 mm at the high elevation head-waters. The region displays a typical nival hydrologic regime with low flows during the snow accumulation period of late autumn to early spring (November to March), and higher spring flows typically starting in April when air temperatures rise above freezing. The Athabasca River is ecologically and economically significant to the development and sustainability of northern Alberta with increasing population and industrial activities, including the multi-billion-dollar oil sands industry [26]. The quantity and quality of flow in the Athabasca River, including extreme high and low flow events, are essential in providing various ecosystem services in the river channels with implication to the downstream Peace Athabasca Delta, which is a UNESCO World Heritage Site and the largest freshwater inland river delta in North America [27].

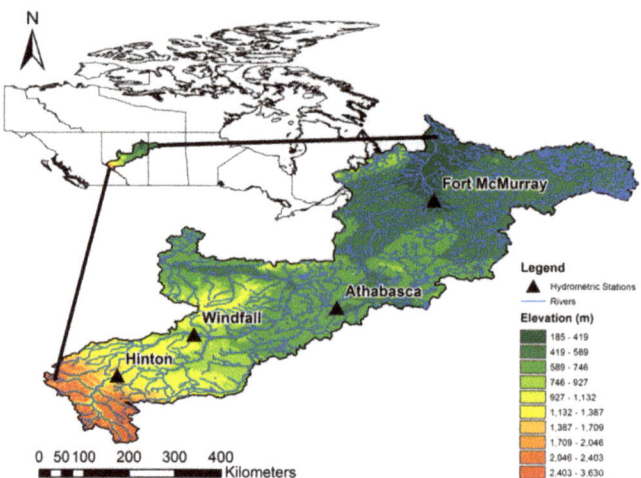

Figure 1. Athabasca watershed with its elevation range and the Athabasca River network, including the locations of the four hydrometric stations used for flood frequency analysis.

2.2. Climate Scenarios and Hydrologic Projections

2.2.1. Climate Model Projection

Regional and local precipitation and potential evaporation are the main climatic drivers controlling the hydrology of a watershed system. A warming climate is shown to affect the magnitude and distribution of both temperature and precipitation that would, in turn, affect the water balance and hydrology of a region [28]. Therefore, studies on the potential impacts of climate change mostly rely on climate projections from global or regional climate models. This study employs statistically downscaled high-resolution gridded daily precipitation, as well as daily maximum (Tmax) and minimum (Tmin) air temperature data to drive a process-based and semi-distributed variable infiltration capacity (VIC) hydrologic model [29] to simulate hydrologic scenarios for the future period. The latest projections originate from twenty-six CMIP5 GCM long-term experimental runs corresponding to the four different levels of representative concentration pathways (RCP2.5, RCP4.5, RCP6.0, and RCP8.5) in which the labels of RCP represent an approximation of the radiative forcing in the year 2100 [30]. Climate projections corresponding to two of the four emission scenarios, namely, the RCP4.5, which is a stabilization scenario that achieve the goal of limiting emission and radiative forcings, and the RCP8.5, which is an emission scenario that greenhouse gas increases as usual until 2100, are selected for hydrologic modelling and analysis in this study. By applying a clustering approach and ranking the models, which differs by region, to provide the widest spread (range) in projected future climate for smaller subsets of the full ensemble, Cannon [31] suggested a set of representative GCMs that fully capture climate variability in 27 extreme climate indices. Moreover, Eum et al. [25] showed that selection of the top six GCMs for Western North America covers over 50% of the variations in the climatic indices considered for the Athabasca River basin. Therefore, the present study uses statistically downscaled data from six GCMs, corresponding to mid-range mitigation (RCP4.5) and high emissions (RCP8.5) scenarios that represent a wider range of climate extremes and seasonal means of precipitation and temperature (see Table 1). Murdock et al. [32] compared the skills of different statistical downscaling (SD) techniques based on sequencing, distribution and spatial pattern related indicators, and recommended two of the more reliable SD techniques, the Bias-Correction Spatial Downscaling (BCSD; [33]) and the bias correction/climate imprint (BCCI; [34]), for regional applications over Canada. The BCSD method uses a quantile-based mapping of the probability density functions for the monthly GCM precipitation and temperature onto those of a gridded observed data spatially aggregated to the GCM scale. Daily results at high spatial resolution are obtained by spatial and temporal disaggregation using rescaled randomly sampled historical observations. The BCCI method uses long-term averages (e.g., 30 years) from the high-resolution observational data as a 'spatial imprint' to represent spatial gradients. The ratio of daily GCM precipitation values to the long-term average monthly climatology of the baseline period is multiplied by the corresponding fine-scale monthly values for a location to get the daily precipitation. These two methods were applied to correct biases and downscale the daily precipitation, Tmax and Tmin scenario data covering the period 1951 to 2100 to a 10-km spatial resolution using the ANUSPLIN observation based gridded data for the 1951–2010 reference period [32]. A total of twenty-four climate projections, from six GCMs and two emission scenarios (RCP4.5 and RCP8.5) and downscaled with two statistical techniques (BCCI and BCSD), are employed to produce an ensemble of hydrologic projections for the Athabasca River basin [25]. This is because future projections by different GCMs usually diverge with time because of different initializations and representations of the various processes in the models and the rate of this divergence is higher for higher emissions scenarios.

Table 1. The select set of six Global Climate Models (GCMs) from the coupled model inter-comparison project (CMIP5) experiment employed in this study.

GCM Abbreviation	Institution	Resolution (Lon. × Lat.)	Primary Reference
CNRM-CM5.1	Centre National de Recherches Meteorologiques and Cerfacs	1.4 × 1.4	Voldoire et al. [35]
CanESM2	Canadian Centre for Climate Modelling and Analysis	2.8 × 2.8	Arora et al. [36]
ACCESS1	Centre for Australian Weather and Climate Research	1.875 × 1.25	Marsland et al. [37]
INM-CM4	Institute of Numerical Mathematics	2.00 × 1.50	Volodin et al. [38]
CSIRO-Mk3.6.0	Commonwealth Scientific and Industrial Research Organisation	1.875 × 1.86	Jeffrey et al. [39]
CCSM4	National Center for Atmospheric Research (NCAR)	1.25 × 0.94	Gent et al. [40]

Figure 2 shows the projected changes in seasonal mean precipitation and air temperature over the Athabasca River basin for the near future (2041–2070 or 2050s) and far future (2071–2100 or 2080s) periods relative to the baseline (1981–2010 or 1990s) period based on those multiple climate projections. The plots indicate overall increases in seasonal precipitation and air temperature over the region except in summer when some models projected decreases in precipitation. In general, the rate of increase in air temperature and precipitation is higher for the higher emission scenario (RCP8.5) compared to the medium RCP4.5 emission scenario and is also higher for the 2080s compared to the 2050s. In particular, there is a strong agreement among all the models with respect to pronounced projected increases in winter air temperature, ranging between 2.5 and 9 $°$C, and precipitation, ranging between 8% and 38% by the end of this century. At the same time, the ranges of climate projections for the RCP8.5/2080s scenario are found to be wider than those for the RCP4.5/2050s indicating that the inter-model variability in the climate projections gets larger with increasing emission concentrations and projection horizons. This is because future projections by different GCMs usually diverge with time because of differences in initial conditions and parameterizations of the various processes in the models and the rate of such divergence gets larger for higher emissions scenarios.

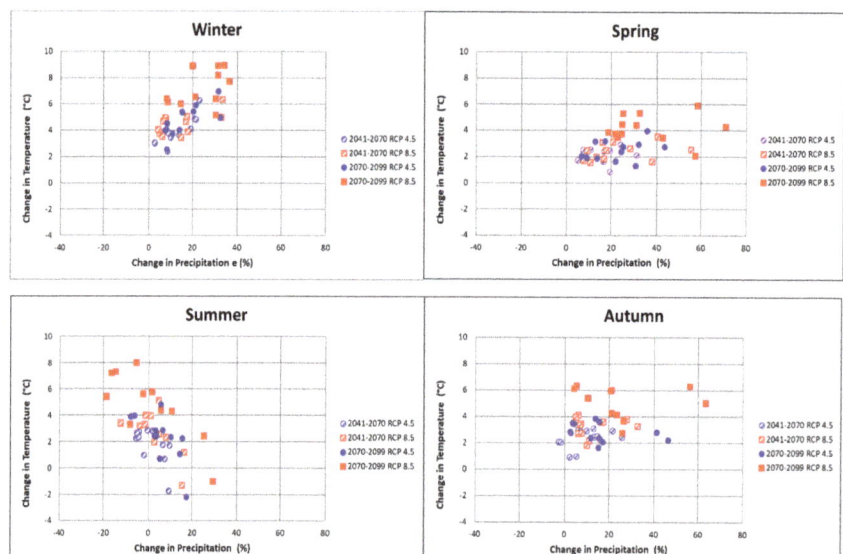

Figure 2. Projected changes in seasonal mean precipitation and air temperature over the Athabasca watershed.

2.2.2. Hydrologic Modelling and River Flow Scenario Simulation

The variable infiltration capacity (VIC), land surface model, is a process-based and spatially distributed macro-scale hydrologic model that simulates the water and energy balances necessary to accurately account for cold-climate hydrologic processes based on prescribed land cover and three-soil layers [29]. The VIC hydrologic model has been successfully applied for evaluating the effects of climate change on hydrologic regimes for watersheds with different basin size, climatology and hydrologic processes [25,41,42]. The model has also been used for evaluation of historical flood events [43] and extreme flow projections [44,45] in several regions. However, one limitation of such hydrologic simulation of flow in cold region rivers is the assumption of open water flow throughout the year and not explicitly accounting for the effect of river-ice freeze-ups, ice-jam and break-up events. Eum et al. [25,46] applied the VIC model over the Athabasca watershed using daily precipitation and temperature data from the ANUSPLIN and statistically downscaled CMIP5 climate model projections. Receiving the daily Tmax, Tmin and Precipitation values, VIC is able to empirically estimate the other energy flux terms over the basin based on geographic coordinates and topographic information. The present study is based on the daily streamflow scenario simulated over the Athabasca watershed by Eum et al. [25] setup of the VIC hydrologic model, with specific emphasis on the analysis of potential changes in peak flows along the Athabasca River, due to projected climate.

The VIC hydrologic model calibration and validation for the Athabasca watershed were performed using daily discharge data at several hydrometric stations along the Athabasca River and its tributaries for the periods 1985–1997 and 1998–2010, respectively [25]. The performance of the calibrated VIC model in replicating the daily mean discharge at four of the hydrometric stations located along the Athabasca River mainstem and that are used for this study is summarized in Table 2. The results show Nash–Sutcliffe (NS) values for the calibration/validation period ranging between 0.78/0.74 and 0.90/0.80. A more detailed description of the VIC model setup used for this study and its calibration/validation results can be found in Eum et al. [25,47].

Table 2. The VIC hydrologic model performances in terms of the Nash–Sutcliffe values during the calibration (1985–1997) and validation (1998–2010) periods.

Station	Hinton	Windfall	Athabasca	Ft.McMurray
Calibration	0.90	0.81	0.78	0.79
Validation	0.78	0.80	0.75	0.74

The calibrated/validated VIC model is applied for hydrologic scenario simulations for the 1990s baseline, as well as for the near (2050s) and far (2080s) future periods, using the twelve sets of statistical downscaled high-resolution climate forcing, corresponding to both the RCP4.5 and RCP8.5 emissions scenarios. Figure 3 presents the box-and-whisker plot of mean monthly streamflow projections and the corresponding changes between the 1990s baseline and the future periods at two locations (a headwater station at Windfall and a downstream station below Fort McMurray) along the Athabasca River. The result indicates an overall projected increase in the Athabasca River discharge for most seasons except in the summer months of July, August, and September that show some decreases. The projected changes are also distinctively higher for the RCP8.5 emissions scenario during the far future period of the 2080s with the increase in mean annual flow ranging between 15.0% to 16.3% of the 1990s baseline value. The corresponding values for the RCP4.5 emissions scenario are relatively smaller, ranging between 7.2% to 11.7% [25].

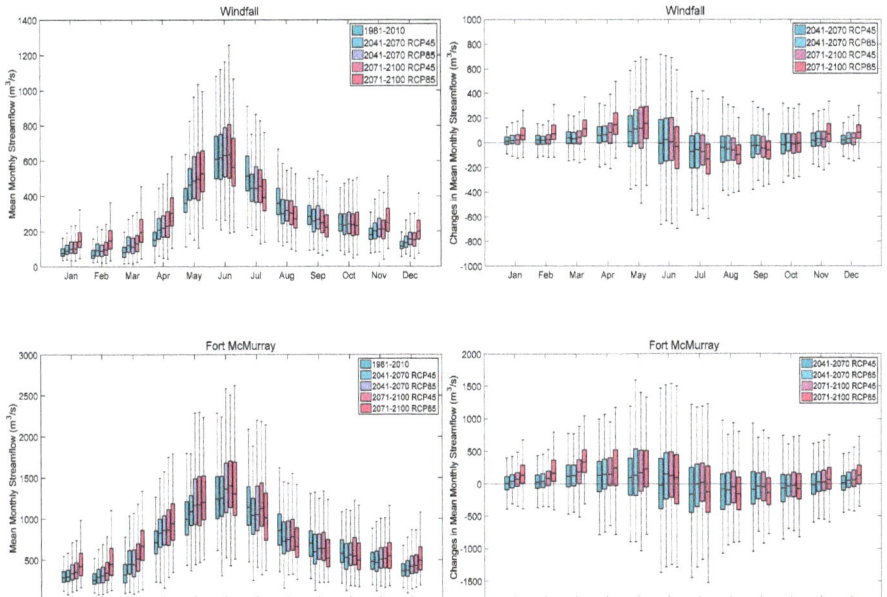

Figure 3. Box-and-Whisker plot of mean monthly streamflow projections and the corresponding changes between the baseline (1990s) and two future periods (2050s and 2080s) at two locations along the Athabasca River based on 12 sets of climate projections (six GCMs × 2 SD) and two RCPs (RCP4.5 and RCP8.5).

2.3. Methods of Peak Flow Analysis

The primary objective of frequency analysis is to relate the magnitude of extreme events to their frequency of occurrence through the use of probability distribution [48]. Two different statistical models of analyzing peak flows, namely, AMS, and partial duration series (PDS) are employed in this study. AMS refers to a series of flow data consisting of the annual maximum daily streamflow values for each year. PDS, on the other hand, includes all independent peak flow events above some pre-defined threshold value. AMS is relatively simpler to apply, as it only requires selecting the annual maximum daily streamflow for analysis; however, some important episodes resulting from multiple independent peak flow events within a water year may be excluded from the study. The advantage of PDS is that it provides the possibility to control the number of flood occurrences to be included in the analysis by appropriate selection of the threshold. However, the choice of threshold and the selection of criteria for retaining flood peaks makes it difficult to use [49]. The specific threshold value for a PDS is usually decided after choosing the average annual number of peak flow events to be included in the PDS. To ensure the selected peak-flow events are independent, inter-event time criteria, specifying the minimum time interval between successive events, and an enter-event discharge level criteria, specifying the minimum flow level between successive events as a fraction of the smaller event, has to be set. After closer examination of the time series data, a minimum inter-even time interval of 72 h and an inter-event level fraction of 0.8 were used to extract the PDS from the daily time series data. This has resulted in 1 to 3 extreme events per year for most of the stations and ensemble members.

The AMS of simulated flows at each of the four hydrometric stations along the Athabasca River main steam corresponding to each of the 12 sets of climate projections (6 GCMs × 2 SD) are extracted for each of the two emission scenarios. A related issue to the magnitude of annual peak flows is the potential shift in the timing of these peak flow events. The Box-and-Whisker plots for the dates of the peak annual flow on Figure 4 show that the median date of AMS in the future scenarios will be getting earlier compared to the baseline period; and more so for the RCP8.5 scenario compared to the RCP4.5. This is an indication that future flooding season will probably shift to an earlier period by order of up to a month or more for the RCP8.5 scenario. Moreover, the variability in the dates of peak flow events will also increase substantially, indicating that the probability of mid-winter and early spring flooding will be increasing.

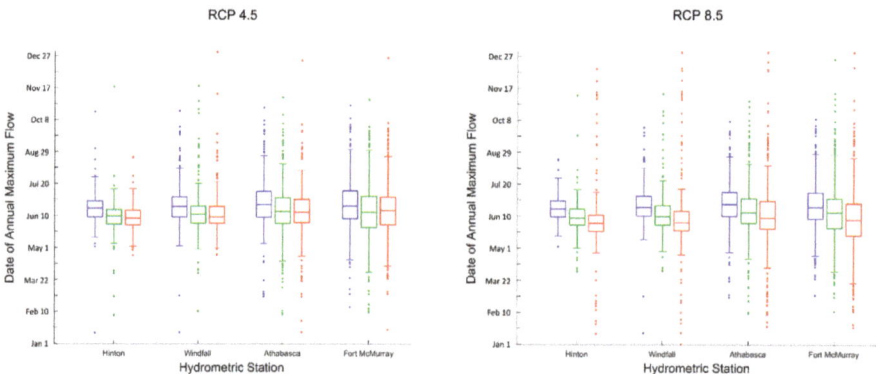

Figure 4. Box-and-Whisker plots of the dates of annual maximum flow series (AMS) corresponding to the two emissions scenario (RCP4.5 and RCP8.5) as simulated at each of the four hydrometric stations along the Athabasca River for the baseline (1990s, blue) and the two future periods (2050s, green) and 2080s (red)).

2.3.1. Stationary and Non-Stationary Analysis

Stationary flood frequency analysis assumes parameters of the probability distribution function for a given period to remain constant. Such analysis is performed in this study on both the AMS and PDS using the Extreme Value Analysis (EVA) tool under the MIKE-Zero platform [50]. Different combinations of six probability distributions (including Gumble, Truncated Gumbel (TGUM), Generalized Extreme Value (GEV), Weibull, Frechet, Log-Pearson Type 3 (LPT3)) and two estimation methods (Method of

Moment (MOM) and Maximum Likelihood (ML)) were evaluated using the standardized least square measure and graphical comparison for fitting the observed peak flow data over the baseline period. While the AMS were best fitted with the LPT3 and the GEV distributions, the PDS were best fitted with LPT3 and TGUM distributions. The parameters for LPT3 distribution were estimated using the MOM, while the parameters for GEV and TGUM distributions were estimated using the ML method. The analysis is done over each of the three 30-year periods (centered at 1990s, 2050s, and 2080s) and the potential impacts of projected climate are estimated by computing the difference in the peak flow magnitudes of various return periods between the baseline and future periods.

However, in a non-stationary world, the probability density functions evolve dynamically over time. Hence, non-stationary analysis works by fitting data to a distribution where the location, scale and shape parameters can be functions of time or climatic variables such as temperature, precipitation or other influencing external factors, such as reservoir operation or land use changes [51]. The non-stationary analysis used in this study applies the generalized additive models for location, scale and shape (or GAMLSS; [52]) on the AMS data. GAMLSS is a general framework for fitting regression-based models that allow all the parameters of the distribution of the response variable to be modelled as linear/non-linear or smooth functions of the explanatory variables. In the present study, the response variable is a series of annual maximum peak discharge that has a parametric cumulative distribution function, and its parameters are modelled as a function of selected covariates. Several distributions under the R package of GAMLSS [53] were tested by modeling the parameters as a linear function of selected covariates and fitting them using the maximum likelihood estimation (MLE) method. Using the Akaike information criterion (AIC), the Schwarz Bayesian criterion (SBC) and by inspecting the quantiles of the residuals, the two parameters—gamma (GA) and log-normal (LNO) distributions—are identified to be most appropriate for the current study. First, time was used as the sole covariate and then annual precipitation and temperature are considered together as alternate covariates. For the latter case, mean annual temperature and annual total precipitation time series over each sub-watershed area contributing to each of the four hydrometric stations are calculated and used as covariates. As an example, Figure 5 presents a non-stationary LNO distribution fitted to simulated peak annual maximum flow series (AMS) at Fort McMurray station with the Log-Normal distribution parameters (μ and δ) varying as a function of time (t). Once the best distribution parameters are fit as functions of the covariates, the projected changes in the frequency and magnitude of peak flow events are computed by averaging their corresponding values over each ten-year period in the 1990s, 2050s, and 2080s. Since there are twelve sets of simulated flow time series (6GCMs and 2DS) corresponding to each of the two emissions scenarios (RCP4.5 & 8.5), the projected changes are mostly reported as the ensemble mean values from all those simulations. The flowchart in Figure 6 illustrates all the different steps and combinations in model simulation, and statistical analysis of peak flows in the Athabasca River.

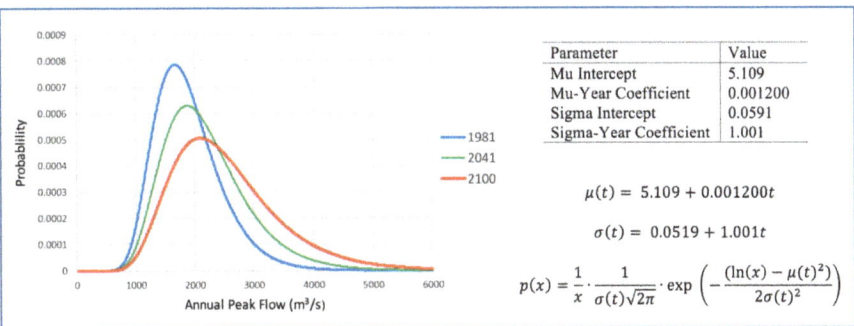

Figure 5. Illustration of a non-stationary LNO distribution fitted to simulated AMS at Fort McMurray station with distribution parameters varying as a function of time (t).

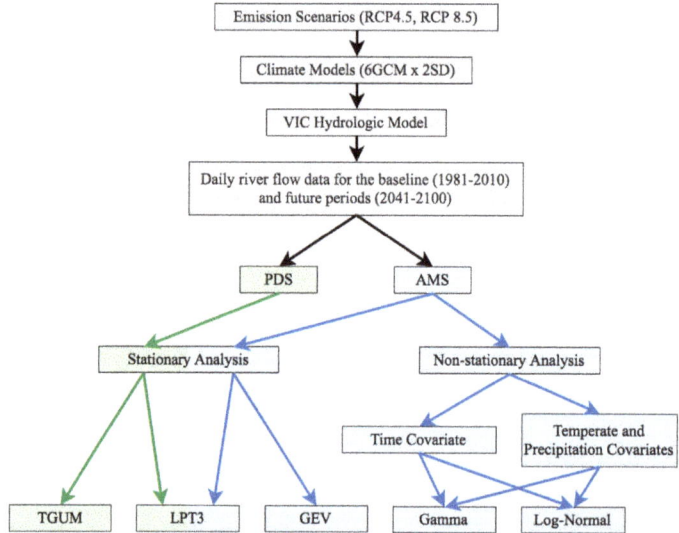

Figure 6. General flowchart showing all the different steps and combinations in hydrologic model simulation and statistical analysis of peak flows in the Athabasca River. GEV, generalized extreme value; PDS, partial duration series; LPT3, log-Pearson type 3; TGUM, Truncated Gumbel.

2.3.2. Uncertainty in Peak Flow Projections

The peak flow analysis in this study employs a set of daily streamflow time series simulated from the VIC hydrologic model of the Athabasca watershed forced with climate data from each of the twelve statistically downscaled GCMs (6GCMs × 2DS). Hence, the projected changes in the frequency of peak flows have a range of possible values resulting from the multiple simulations. In addition, both stationary and non-stationary analysis techniques are applied to each simulated streamflow time series, with different distribution functions and covariates, resulting in eight sets of outcomes for each streamflow projection. The sensitivity of projected changes in the frequency of peak flows to the driving climate models, as well as the statistical methods of analysis is examined by calculating the inter-climate model variability and the inter-statistical method variability in terms of their corresponding standard deviations. While inter-climate model standard deviation for each statistical method of frequency analysis is calculated from multiple projected changes corresponding to each climate models, inter-statistical model standard deviations corresponding to each driving climate model are calculated from projected changes by multiple statistical methods of analysis. Finally, the sensitivity of the projected change in peak flow to inter-climate model variability is compared with that of the inter-statistical method variability.

3. Results

3.1. Stationary Analysis

The stationary analysis techniques are applied on both the AMS and the PDS, at each of the four hydrometric stations along the Athabasca River. The time series of peak flow is derived from the VIC simulated streamflow data corresponding to each GCM, statistical downscaling (SD) methods and emissions scenario combination (RCPs). Analysis results are presented as peak flow magnitudes and corresponding changes for a number of events between 2- and a 100-year return periods. The results are then averaged over all the driving GCM/SD to create ensemble mean values for each future period and emissions scenario combination. Figure 7 shows the ensemble mean projected changes in peak flow events of different return periods between the 1990s baseline and the 2080s future periods corresponding to the RCP8.5 emissions scenario. The results indicate an overall decrease in the return period (or increase in the frequency) of most flow quantiles for the future period. At Fort McMurray, for instance, a 100-year peak flow event for the baseline period will become a 30-year one by the end of this century and a 50-year peak flow event will become more frequently than once in 20 years. Moreover, the average magnitudes of changes are relatively lower for the headwater stations at Hinton and Windfall compared to the downstream stations at Athabasca and Fort McMurray. Moreover, the ranges of predicted changes for the upstream stations are generally wider than those for the downstream stations indicating the increased uncertainty of the results for the upstream stations. Potential changes in the frequency of peak flow magnitudes, estimated using the AMS series are generally higher than those using the PDS. However, there seems to be no consistent pattern in the projected changes that can be attributable to the specific statistical methods applied to model the frequency distributions. The results corresponding to the RCP4.5 emission scenario (not presented) are very similar to that of the RCP8.5 except that the changes are relatively smaller for the former, with the 100-year peak flow at Fort McMurray becoming a 50-year one and a 50-year peak flow becoming a 30-year one by the end of the century.

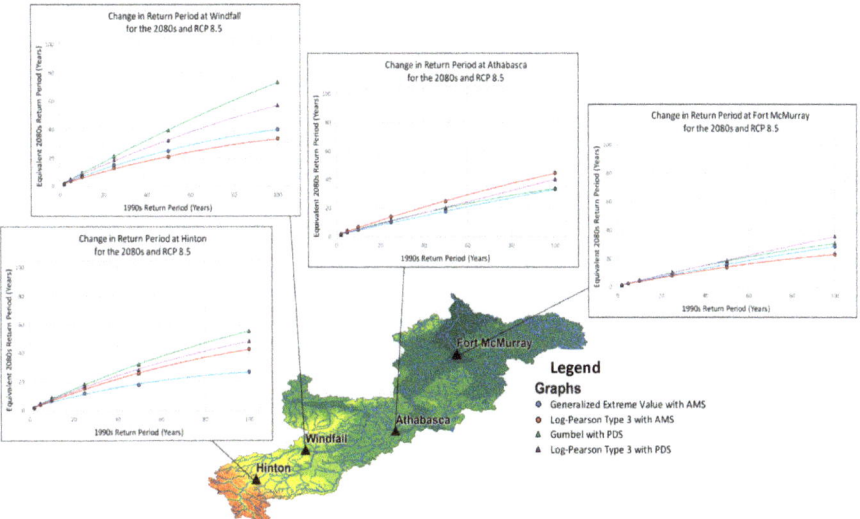

Figure 7. Results of stationary analysis - ensemble mean projected changes in flood events of different return periods between the 1990s baseline and the 2080s future periods corresponding to the RCP8.5 emissions scenario.

3.2. Non-Stationary Analysis

Non-stationary analysis on the AMS is performed first using time (as the number of years from the start of the AMS; i.e., 1981) as the only covariate and then using both the mean annual temperature and annual total precipitation as covariates on which the values of the distribution parameters for the Gamma and Log-Normal distributions depend. For each of the four hydrometric stations considered, the mean temperature and precipitation covariates are computed only over the region contributing (draining) to each of the measuring stations. As in Figures 7 and 8 shows the ensemble mean projected changes in peak flow events of different return periods between the 1990s baseline and the 2080s future periods corresponding to the RCP8.5 emissions scenario. The magnitude and direction of changes in the ensemble mean results from the non-stationary analysis are generally similar to those of the stationary analysis for the two headwater stations (Hinton and Windfall) except that the ranges of the projected changes are narrower for the former. For the two downstream stations, however, the non-stationary analysis predicted greater changes (decreases) in the return periods of low frequency events than that of the stationary approach.

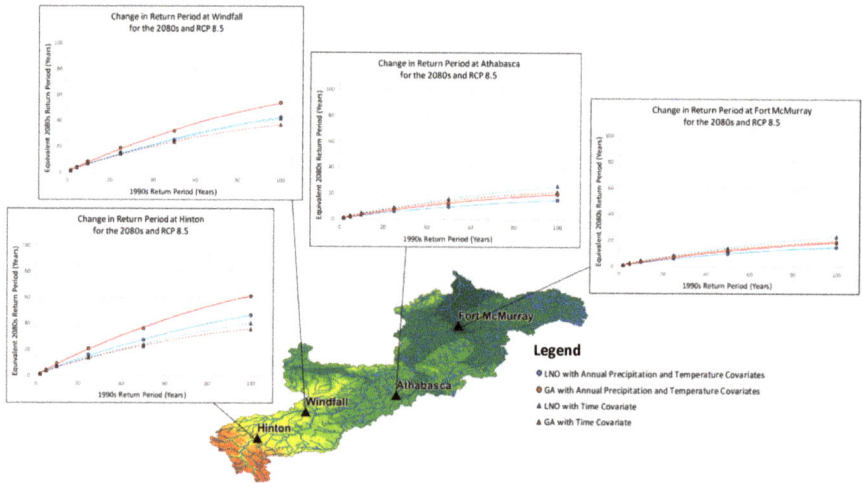

Figure 8. Results of non-stationary analysis - ensemble projected changes in flood events of different return periods corresponding to the 1990s baseline and the 2080s future periods corresponding to the RCP8.5 emissions scenarios. The analysis is conducted on AMS using time, as well as temperature and precipitation covariates.

Moreover, using time as the only covariate predicted larger decreases in the return periods compared to using mean annual temperature and precipitation as covariates for the two upstream stations, while the reverse is true for the remaining two downstream stations (Athabasca and Fort McMurray). For example, a 100-year peak flow event for the baseline period will become a 35- to 62- year event at Hinton headwater station or a 15- to 22-year event at the downstream Fort McMurray station by the end of this century. When using precipitation and temperature as covariates, the Log-Normal distribution resulted in greater projected changes (decreases) in return periods compared to the Gama distribution. On the contrary, when using only time as a covariate, the Log-Normal distribution resulted in smaller projected changes (decreases) in return periods compared to the Gama distribution. Consistent with the case of the stationary analysis, the ranges of predicted changes for the upstream stations are generally wider than those for the downstream stations, again indicating the higher uncertainty in the results for the upstream stations.

3.3. Changes in Peak Flows

Figure 9 presents the ensemble mean projected changes (%) in the magnitude of peak flow events of different return periods between the 2080s and the 1990s baseline period for the RCP8.5 emissions scenario based on both stationary and non-stationary methods of frequency analysis. The percentage of projected changes at each location varies depending on the statistical method of analysis, such as stationary vs non-stationary analysis, type of distribution function applied and the covariates used to model the parameters. The changes in the peak flow magnitudes generally get larger with increases in the return period. Relative changes at the downstream stations (Athabasca and Fort McMurray) are also generally higher than those at the headwater stations (Hinton and Windfall) resulting from accumulated effects of increasing flows within the drain area from the headwater to downstream stations. To show a specific example, peak flow events with a 100-year return period at the two upstream stations is projected to increases by about 4% to 12% during 2080s compared to the 1990s baseline period. The corresponding increases for the two downstream stations range from 21 to 33%. At the same time, the corresponding increases in peak flow events with 5-year return period at the two upstream stations vary from 1% to 9%, while it varies from 14% to 25% at the two downstream stations. Similar increases in peak flow magnitude are projected for the RCP4.5 emissions scenarios; however, the changes are slightly smaller with projected increases in the 100-year peak flow at the two downstream stations varying from 18% to 28% (not shown).

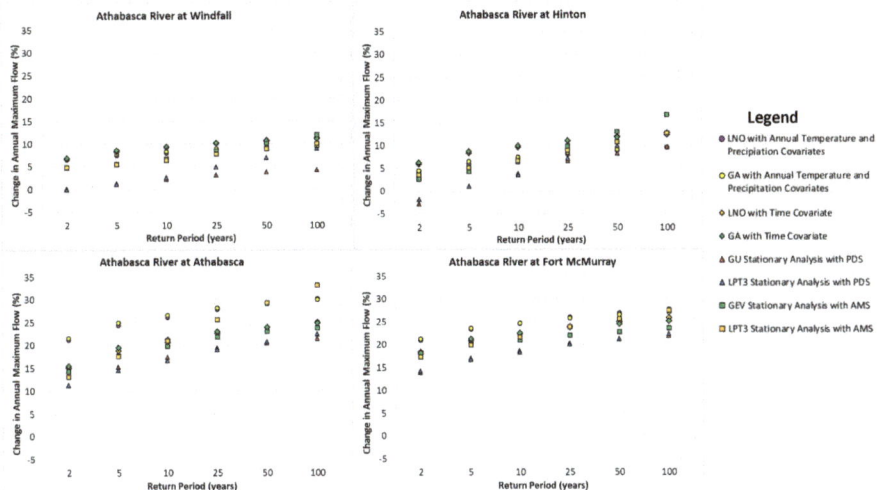

Figure 9. Ensemble mean projected changes (%) in the magnitude of flood events of different return periods between the 2080s and the 1990s baseline period for the RCP8.5 emissions scenario corresponding to both stationary and non-stationary methods of extreme flow analysis.

The non-stationary analysis mostly results in greater projected changes in peak flows than the stationary ones. The smallest relative changes resulted from the stationary analysis performed on the PDS data, while non-stationary analysis with AMS and using precipitation and temperature as covariate resulted in the biggest projected changes. For example, while stationary analysis performed on the PDS at each of the four stations resulted in 4% to 23% increases in the 100-year peak flow event by the 2080s, compared to the 1990s, the corresponding increase for non-stationary analysis with AMS using precipitation and temperature as covariate range between 10% to 30%.

3.4. Inter-Model Variability

The analysis results presented above show that projected changes in the magnitude and frequency of peak flow results for a given emission scenario and future horizon depends on both the climate models corresponding to the streamflow projection and the statistical method of frequency analysis. Inter-climate model variability of projected changes in peak flow corresponding to each method of extreme flow analysis is presented in Figure 10. The result shows the inter-climate model standard deviation of the changes between the 2080s and the 1990s baseline period and the RCP8.5 emissions scenario for a range of return periods. The inter-climate model variability in projected changes is generally larger as the return period gets longer. For example, while the ensemble mean value of projected changes in the 100-year peak flow at the Fort McMurray station range between 22% and 28%, its standard deviation range between 19% and 39% depending on the statistical method of analysis. While the stationary analysis with AMS produces the greatest inter-climate model variability, the non-stationary analysis with precipitation and temperature co-variate generally produces the smallest inter-climate model variability. This is an indication that the uncertainty in the parameters of the frequency distributions is reduced by using the driving temperature and precipitation as co-variates to constrain their values. There seems to be no systematic difference in the pattern of inter-climate model variability between the different stations.

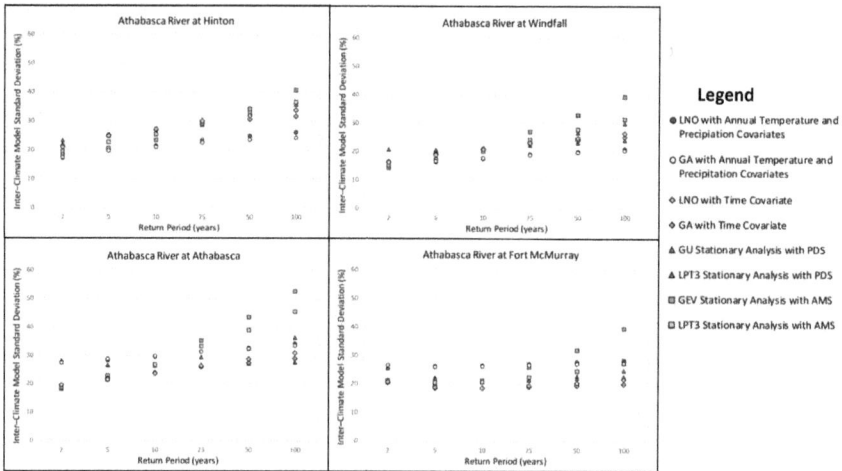

Figure 10. Inter- climate model variability of projected changes in peak flow between the 2080s and the 1990s baseline period for the RCP8.5 emissions scenario corresponding to each of the statistical methods considered.

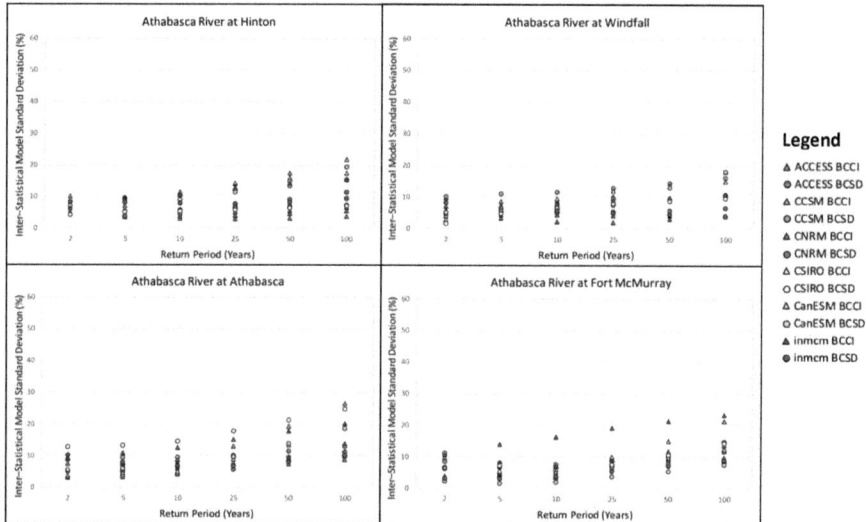

Figure 11. Inter-statistical model variability of projected changes in peak flow between the 2080s and the 1990s baseline period and the RCP8.5 emissions scenario corresponding to each of the GCM/SD considered.

Figure 11 presents the inter-statistical model variability of projected changes in peak flow between the 2080s and the 1990s baseline period and the RCP8.5 emissions scenario for each of the climate models considered. The inter-statistical model standard deviation of projected changes shows a similar pattern at all the four stations with a gradual increase with an increase in the return period. For example, the inter-statistical model variation in the 100-year peak flow at the Fort McMurray station ranges between 7% and 24% depending on the climate model employed. While there are different patterns of inter-statistical model variabilities corresponding to each of the climate models, the ranges of variabilities are very similar with slightly higher values and wider ranges for higher return periods. However, when compared to the inter-climate model standard deviation, the inter-statistical model standard deviations are generally smaller, indicating that the uncertainty in projected changes resulting from the driving climate models is generally higher than that coming from the statistical methods of extreme flow analysis.

4. Summary and Conclusions

The study examines potential changes in the frequency and magnitude of peak flows in the Athabasca River in Alberta, Canada based on simulated discharges for the future climate. The daily stream flows for the baseline and future periods are simulated with the VIC hydrological model of the Athabasca watershed driven by multiple statistically-downscaled high-resolution climate scenarios corresponding to the RCP4.5 and RCP8.5 emissions scenarios. Analysis of simulated flows generally indicates potential increases during the winter and spring and decreases during the summer and early fall seasons, with an overall increase in high flows, especially for low frequency peak flow events. However, the study also reveals that projected changes in the frequency and magnitude of peak flow events vary over a wide range, especially for low frequency events, depending on the climate model/data used to simulate the streamflow and the statistical method of peak flow analysis. For example, the ensemble mean projected changes for the 100-year peak flow event by 2080s ranges from 4% to 33% depending on emissions scenarios and the statistical method of analysis. These increases correspond to a 100-year peak flow event of the 1990s baseline period becoming a 20- to 50-year event at the end of the current century with larger changes at downstream stations compared to upstream ones. While all peak flows may not necessarily cause flooding, the projected increase in the frequency and magnitude of future peak flow events are most likely to increase the probability of flooding in specific reaches (with floodplain) along the river.

The non-stationary peak flow analyses show relatively larger increases in peak flow magnitudes at different return periods compared to that of the stationary methods, especially for the downstream stations. The stationary analysis on the PDS resulted in smaller projected changes in peak flows than that of the AMS. However, the application of stationary analysis over multiple 30-years epochs as compared to a combined 90 years of data for the non-stationary analysis may have some bearing on the comparison two approaches. The two non-stationary approaches, one using time as the only covariate, and the other using precipitation and temperature as covariates, have also produced slightly different results that can be explained by the nature of the covariates. With only time as a covariate, the changes in the model parameters are linear, while using the temperature and precipitation covariates, the changes in the model parameters are non-linear as they depend on the variation in temperature and precipitation. This seems to allow the multivariate analysis, with temperature and precipitation as covariates, to fit better to the changing frequency of peak flows. The effect of the distribution applied (log-normal vs gamma) on the magnitude of the changes is also found to be different depending on the covariates employed and no specific distribution seems to produce the consistently higher or lower magnitude of changes for all the different cases. The study also showed that inter-model variabilities generally increase with increases in the return periods, mostly because there is an increasing reliance on distribution characteristics for predicting less frequent events. In general, the projected changes in the frequency and magnitude of peak flow events vary depending on both the driving GCMs and the statistical methods of peak flow analysis. However, the sensitivity of changes to the statistical method of analysis is generally smaller compared to that resulting from inter-climate model variability. Therefore, while the issue of non-stationarity is important in future peak flow projection, considering the range of model projections for the future climate condition is equally or even more important.

Author Contributions: Conceptualization, Y.D., H-I.E. and P.C.; methodology, Y.D., H-I.E. and J.H..; investigation and analysis, Y.D and J.H.; writing—original draft preparation, Y.D and J.H.; writing—review and editing, P.C. and H-I.E.; project administration, Y.D.; funding acquisition, Y.D."

Funding: This study was conducted with the financial support provided by the Environment and Climate Change Canada's Climate Change Adaptation program and the Joint Oil-Sands Monitoring Program (JOSMP).

Acknowledgments: This project was conducted in collaboration with the NSERC funded Canadian FloodNet project. The authors also acknowledge the contribution of Émilie Wong and Victoria Gagnon at the various stages of data acquisition and processing.

Conflicts of Interest: The authors declare no conflict of interest.

References

1. Cunderlik, J.M.; Ouarda, T.B.M.J. Trends in the timing and magnitude of floods in Canada. *J. Hydrol.* **2009**, *375*, 471–480. [CrossRef]
2. Park, D.; Markus, M. Analysis of a changing hydrologic flood regime using the Variable Infiltration Capacity model. *J. Hydrol.* **2014**, *515*, 267–280. [CrossRef]
3. Buttle, J.M.; Allen, D.M.; Caissie, D.; Davison, B.; Hayashi, M.; Peters, D.L.; Pomeroy, J.W.; Simonovic, S.; St-Hilaire, A.; Whitfield, P.H. Flood processes in Canada: regional and special aspects. *Can. Water Resour. J.* **2016**, *41*, 7–30. [CrossRef]
4. Beltaos, S.; Prowse, T. River-ice hydrology in a shrinking cryosphere. *Hydrol. Process.* **2009**, *23*, 122–144. [CrossRef]
5. Prowse, T.D.; Beltaos, S. Climatic control of river-ice hydrology: a review. *Hydrol. Process.* **2002**, *16*, 805–822. [CrossRef]
6. Burn, D.H.; Sharif, M.; Zhang, K. Detection of trends in hydrological extremes for Canadian watersheds. *Hydrol. Process.* **2010**, *24*, 1781–1790. [CrossRef]
7. Ganguli, P.; Coulibaly, P. Does nonstationarity in rainfall require nonstationary intensity–duration–frequency curves? *Hydrol. Earth Syst. Sci.* **2017**, *21*, 6461–6483. [CrossRef]
8. Burn, D.H. Climatic influences on streamflow timing in the headwaters of the Mackenzie River Basin. *J. Hydrol.* **2008**, *352*, 225–238. [CrossRef]
9. Rao, A.R.; Hamed, K.H. *Flood Frequency Analysis*; CRC press: New York, NY, USA, 2000.
10. Cooley, D. Return periods and return levels under climate change. In *Extremes in a Changing Climate*; AghaKouchak, A., Easterling, D., Hsu, K., Sorooshian, S., Eds.; Springer: Dordrecht, The Netherlands, 2013; Volume 65, pp. 97–114.
11. Tramblay, Y.; Neppel, L.; Carreau, J.; Najib, K. Non-stationary frequency analysis of heavy rainfall events in southern France. *Hydrol. Sci. J.* **2013**, *58*, 280–294. [CrossRef]
12. Khaliq, M.N.; Ouarda, T.B.M.J.; Ondo, J.C.; Gachon, P.; Bob´ee, B. Frequency analysis of a sequence of dependent and/or non-stationary hydro-meteorological observations: A review. *J. Hydrol.* **2006**, *329*, 534–552. [CrossRef]
13. Cunderlik, J.M.; Burn, D.H. Non-stationary pooled flood frequency analysis. *J. Hydrol.* **2003**, *276*, 210–223. [CrossRef]
14. Tan, X.; Gan, T.Y. Nonstationary analysis of annual maximum streamflow of Canada. *J. Clim.* **2015**, *28*, 1788–1805. [CrossRef]
15. Ouarda, T.B.M.J.; El-Adlouni, S. Bayesian nonstationary frequency analysis of hydrological variables 1. *JAWRA* **2011**, *47*, 496–505.
16. López, J.; Francés, F. Non-stationary flood frequency analysis in continental Spanish rivers, using climate and reservoir indices as external covariates. *Hydrol. Earth Syst. Sci.* **2013**, *17*, 3189–3203. [CrossRef]
17. Li, J.; Tan, S. Nonstationary flood frequency analysis for annual flood peak series, adopting climate indices and check dam index as covariates. *Water Resour. Manag.* **2015**, *29*, 5533–5550. [CrossRef]
18. Seidou, O.; Ramsay, A.; Nistor, I. Climate change impacts on extreme floods I: combining imperfect deterministic simulations and non-stationary frequency analysis. *Nat. Hazards* **2012**, *61*, 647–659. [CrossRef]
19. Zhang, T.; Wang, Y.; Wang, B.; Tan, S.; Feng, P. Nonstationary flood frequency analysis using univariate and bivariate time-varying models based on GAMLSS. *Water* **2018**, *10*, 819. [CrossRef]

20. Dong, N.D.; Agilan, V.; Jayakumar, K.V. Bivariate Flood Frequency Analysis of Nonstationary Flood Characteristics. *J. Hydrol. Eng.* **2019**, *24*, 04019007. [CrossRef]
21. Shrestha, R.R.; Cannon, A.J.; Schnorbus, M.A.; Zwiers, F.W. Projecting future nonstationary extreme streamflow for the Fraser River, Canada. *Clim. Chang.* **2017**, *145*, 289–303. [CrossRef]
22. Bonsal, B.R.; Cuell, C. Hydro-climatic variability and extremes over the Athabasca River basin: Historical trends and projected future occurrence. *Can. Water Resour. J.* **2017**, *42*, 315–335. [CrossRef]
23. Dibike, Y.; Prowse, T.; Bonsal, B.; O'Neil, H. Implications of future climate on water availability in the western Canadian river basins. *Int. J. Clim.* **2017**, *37*, 3247–3263. [CrossRef]
24. Dibike, Y.; Eum, H.I.; Prowse, T. Modelling the Athabasca watershed snow response to a changing climate. *J. Hydrol. Reg. Stud.* **2018**, *15*, 134–148. [CrossRef]
25. Eum, H.I.; Dibike, Y.; Prowse, T. Climate-induced alteration of hydrologic indicators in the Athabasca River Basin, Alberta, Canada. *J. Hydrol.* **2017**, *544*, 327–342. [CrossRef]
26. Rogers, M.E. *Surface Oil Sands Water Management Summary Report*; Cumulative Environmental Management Association (CEMA): Fort McMurray, AB, Canada, 2010.
27. Prowse, T.D.; Beltaos, S.; Gardner, J.T.; Gibson, J.J.; Granger, R.J.; Leconte, R.; Peters, D.L.; Pietroniro, A.; Romolo, L.A.; Toth, B. Climate change, flow regulation and land-use effects on the hydrology of the Peace-Athabasca-Slave system; Findings from the Northern Rivers Ecosystem Initiative. *Environ. Monit. Assess.* **2006**, *113*, 167–197. [CrossRef] [PubMed]
28. Stocker, T.; Qin, D. (Eds.) *Climate Change 2013: The Physical Science Basis: Working Group I Contribution to the Fifth Assessment Report of the Intergovernmental Panel on Climate Change*; Cambridge University Press: New York, NY, USA, 2014.
29. Liang, X. A Two-Layer Variable Infiltration Capacity Land Surface Representation for General Circulation Models. Ph.D. Dissertation, NASA, Washington, DC, USA, 1994.
30. Taylor, K.E.; Stouffer, R.J.; Meehl, G.A. An overview of CMIP5 and the experiment design. *Bull. Am. Meteorol. Soc.* **2012**, *93*, 485–498. [CrossRef]
31. Cannon, A.J. Selecting GCM scenarios that span the range of changes in a multimodel ensemble: application to CMIP5 climate extremes indices. *J. Clim.* **2015**, *28*, 1260–1267. [CrossRef]
32. Murdock, T.Q.; Cannon, A.J.; Sobie, S.R. *Statistical Downscaling of Future Climate Projections*; Pacific Climate Impacts Consortium (PCIC): Victoria, BC, Canada, 2013.
33. Maurer, E.P.; Hidalgo, H.G. Utility of daily vs. monthly large-scale climate data: an intercomparison of two statistical downscaling methods. *Hydrol. Earth Syst. Sci.* **2008**, *12*, 551–563. [CrossRef]
34. Hunter, R.D.; Meentemeyer, R.K. Climatologically aided mapping of daily precipitation and temperature. *J. Appl. Meteorol.* **2005**, *44*, 1501–1510. [CrossRef]
35. Voldoire, A.; Sanchez-Gomez, E.; Mélia, D.S.; Decharme, B.; Cassou, C.; Sénési, S.; Valcke, S.; Beau, I.; Alias, A.; Chevallier, M.; et al. The CNRM-CM5.1 global climate model: description and basic evaluation. *Clim. Dyn.* **2013**, *40*, 2091–2121. [CrossRef]
36. Arora, V.K.; Scinocca, J.F.; Boer, G.J.; Christian, J.R.; Denman, K.L.; Flato, G.M.; Kharin, V.V.; Lee, W.S.; Merryfield, W.J. Carbon emission limits required to satisfy future representative concentration pathway of greenhouse gases. *Geophys. Res. Lett.* **2011**, *38*, L05805. [CrossRef]
37. Marsland, S.J.; Bi, D.; Uotila, P.; Fiedler, R.; Griffies, S.M.; Lorbacher, K.; O'Farrell, S.; Sullivan, A.; Uhe, P.; Zhou, X.; et al. Evaluation of ACCESS climate model ocean diagnostics in CMIP5 simulations. *Austral. Meteorol. Oceanogr. J.* **2013**, *63*, 101–119. [CrossRef]
38. Volodin, E.M.; Dianskii, N.A.; Gusev, A.V. Simulating present-day climate with the INMCM4.0 coupled model of the atmospheric and oceanic general circulations. *Izv. Atmos. Ocean. Phys.* **2010**, *46*, 414–431. [CrossRef]
39. Jeffrey, S.J.; Rotstayn, L.D.; Collier, M.A.; Dravitzki, S.M.; Hamalainen, C.; Moeseneder, C.; Wong, K.K.; Syktus, J.I. Australia's CMIP5 submission using the CSIRO Mk3.6 model. *Austral. Meteorol. Oceanogr. J.* **2013**, *63*, 1–13. [CrossRef]
40. Gent, P.R.; Danabasoglu, G.; Donner, L.J.; Holland, M.M.; Hunke, E.C.; Jayne, S.R.; Lawrence, D.M.; Neale, R.B.; Rasch, P.J.; Vertenstein, M.; et al. The community climate system model version 4. *J. Clim.* **2011**, *24*, 4973–4991. [CrossRef]

41. Werner, A.T.; Schnorbus, M.A.; Shrestha, R.R.; Eckstrand, H.D. Spatial and temporal change in the hydro-climatology of the Canadian portion of the Columbia River basin under multiple emissions scenarios. *Atmos. Ocean.* **2013**, *51*, 357–379. [CrossRef]
42. Elsner, M.M.; Cuo, L.; Voisin, N.; Deems, J.S.; Hamlet, A.F.; Vano, J.A.; Mickelson, K.E.; Lee, S.Y.; Lettenmaier, D.P. Implications of 21st century climate change for the hydrology of Washington State. *Clim. Chang.* **2010**, *102*, 225–260. [CrossRef]
43. Wenger, S.J.; Luce, C.H.; Hamlet, A.F.; Isaak, D.J.; Neville, H.M. Macroscale hydrologic modeling of ecologically relevant flow metrics. *Water Resour. Res.* **2010**, *46*, W09513. [CrossRef]
44. Västilä, K.; Kummu, M.; Sangmanee, C.; Chinvanno, S. Modelling climate change impacts on the flood pulse in the Lower Mekong floodplains. *J. Water Clim. Chang.* **2010**, *1*, 67–86. [CrossRef]
45. Hamlet, A.F.; Lettenmaier, D.P. Effects of 20th century warming and climate variability on flood risk in the western US. *Water Resour. Res.* **2007**, *43*. [CrossRef]
46. Eum, H.I.; Dibike, Y.; Prowse, T. Comparative evaluation of the effects of climate and land-cover changes on hydrologic responses of the Muskeg River, Alberta, Canada. *J. Hydrol. Reg. Stud.* **2016**, *8*, 198–221. [CrossRef]
47. Eum, H.I.; Yonas, D.; Prowse, T. Uncertainty in modelling the hydrologic responses of a large watershed: a case study of the Athabasca River Basin, Canada. *Hydrol. Process.* **2014**, *28*, 4272–4293. [CrossRef]
48. Chow, V.T.; Maidment, D.R.; Mays, L.W. *Applied Hydrology*; Editions McGraw-Hill: New York, NY, USA, 1988; 572p.
49. Lang, M.; Ouarda, T.B.M.J.; Bobée, B. Towards operational guidelines for over-threshold modeling. *J. Hydrol.* **1999**, *225*, 103–117. [CrossRef]
50. DHI. Extreme Value Analysis (EVA) Technical Reference and Documentation. 2017. Available online: manuals.mikepoweredbydhi.help/2017/General/EVA_SciDoc.pdf (accessed on 12 June 2019).
51. Salvadori, G.; De Michele, C. Multivariate extreme value methods. In *Extreme in a Changing Climate*; Springer: Dordrecht, The Netherlands, 2013; pp. 115–162.
52. Rigby, R.A.; Stasinopoulos, D.M. Generalized additive models for location, scale and shape. *J. R. Stat. Soc. Ser. C (Appl. Stat.)* **2005**, *54*, 507–554. [CrossRef]
53. Stasinopoulos, M.D.; Rigby, R.A.; Heller, G.Z.; Voudouris, V.; De Bastiani, F. *Flexible Regression and Smoothing: Using GAMLSS in R*; Chapman and Hall/CRC: Boca Raton, FL, USA, 2017.

© 2019 by the authors. Licensee MDPI, Basel, Switzerland. This article is an open access article distributed under the terms and conditions of the Creative Commons Attribution (CC BY) license (http://creativecommons.org/licenses/by/4.0/).

Article

Climate Change Induced Salinization of Drinking Water Inlets along a Tidal Branch of the Rhine River: Impact Assessment and an Adaptive Strategy for Water Resources Management

Matthijs van den Brink [1,*], Ymkje Huismans [2], Meinte Blaas [3] and Gertjan Zwolsman [4]

1. HydroLogic, P.O. Box 2177, 3800 CD Amersfoort, The Netherlands
2. Deltares (Unit Marine and Coastal Systems), P.O. Box 177, 2600 MH Delft, The Netherlands; ymkje.huismans@deltares.nl
3. Rijkswaterstaat (Unit of Water, Traffic & Environment, Dept. of Water Management), Ministry of Infrastructure & Water Management, P.O. Box 2232, 3500 GE Utrecht, The Netherlands; meinte.blaas@rws.nl
4. Dunea, Plein van de Verenigde Naties 11, 2719 EG Zoetermeer, The Netherlands; g.zwolsman@dunea.nl
* Correspondence: matthijs.vandenbrink@hydrologic.com; Tel.: +31-6-50636712

Received: 28 February 2019; Accepted: 26 March 2019; Published: 2 April 2019

Abstract: This study presents the results of an impact analysis of climate change on salinization and the long-term availability of drinking water resources along the river Lek, a tidal branch of the Rhine delta, and a potential mitigation measure. To this end, a one-dimensional modelling approach was used that enabled studying 50 years of variation in discharge and tide in current and future climate. It was found that all locations are increasingly vulnerable to salt intrusion caused by the combination of sea level rise and decreasing river discharges. This affects both the yearly average chloride concentration and long duration exceedances of the threshold value of 150 mg/L. It was also found that diverting a higher fresh water discharge to the Lek of several tens of cubic meters per second reduces the risk of salinization at the upstream inlet locations. However, the increased influence of seawater intrusion on the drinking water inlets cannot be fully compensated for by this measure. The potential gain of the extra water for the drinking water inlets along the Lek has to be balanced against the impact of this measure on water levels and stream flows in other parts of the river system.

Keywords: climate change; salinization; water resources management; drinking water

1. Introduction

The Netherlands constitute a densely populated part of the Rhine–Meuse delta. Due to an annual precipitation surplus of 300 mm and border crossing discharges of the Rhine and Meuse rivers (averaging at 2200 m^3/s and 230 m^3/s respectively), fresh water supply has not been an issue before long. However, the country's limited elevation and the proximity of the sea make The Netherlands vulnerable to seawater intrusion and salinization of freshwater inlets.

Seawater intrusion in river deltas is largely governed by two variables, both subject to climate change:

- Sea level rise. Climate projections for The Netherlands show an estimated sea level rise of 0.15 to 0.40 m in the year 2050, and an increase of 0.25 to 0.80 m by 2085, compared to the reference year 1985 [1].
- A lower river baseflow. The regionalized climate projections indicate a potential worst case decrease of 20 percent of the annual 7-day minimum discharge for the Rhine in 2050 and a 30 percent decrease in 2085, compared to the reference year 1985 [2].

Both drivers cause the sea water to penetrate further inland, through the open river–sea connection in the estuary.

Several impact assessments [3,4] show that in the long term, fresh water supply in The Netherlands is at risk, especially in the low-lying western area where salt water intrusion occurs, whereas the dependence on surface water is highest, due to the presence of brackish ground water. As part of the national Delta Programme [5], national and regional governments, water authorities and public and private water users jointly seek opportunities to make fresh water supply resilient to climate change. This study presents the results of an impact analysis of climate change on the long-term availability of drinking water resources in the river Lek, a tidal branch of the Rhine delta. In addition, we will explore a potential measure to protect these freshwater resources by reallocating the Rhine discharge over its various branches in the delta.

The Lek serves as a drinking water source to approx. 2.2 million inhabitants in the southwest The Netherlands. There are six indirect abstractions present along the river (river bank filtration) and one direct surface water intake. The water treatment does not include desalination, as the chloride concentration of the river water hardly ever exceeds the drinking water standard of 150 mg/L. However, given the impacts of climate change on river flow and sea level rise, it is conceivable that some of the freshwater intakes along the Lek are vulnerable to salinization.

The research questions of the analysis are:

1. To what extent may climate change increase the probability of salt intrusion on the Lek, limiting its quality as drinking water source?
2. To what extent can salt water intrusion be reduced by diverting a higher fresh water discharge towards the Lek?

These research questions are addressed by this study using a mathematical modelling approach, integrating river flow, seawater level and salt loads of the system. As salt intrusion on the Lek has not been studied in such detail before, an existing salt transport model of the estuary was updated in order to assess the vulnerability of the freshwater inlets along the Lek towards salinization due to climate change. Additionally, the model was used to assess the effectiveness of diverting higher freshwater discharges to the Lek, in order to alleviate future salinization events. The effectiveness of the diversion strategy is evaluated in terms of our understanding of when and where different salinization mechanisms prevail.

2. System Description, Methods and Materials

2.1. Northern Rhine Delta Basin (NDB) System Description

The Rhine river splits into several branches just upstream of the city of Arnhem (Figure 1). The distribution of the discharge over the branches can be controlled to a limited extent by a weir near the city of Arnhem. On average, two thirds of the river flow is directed to the Waal branch, while the Lek receives some 10–15% of the cross border flow of the Rhine. The allocation of the river flow along the Lek river can be controlled by three weirs (see Figure 1).

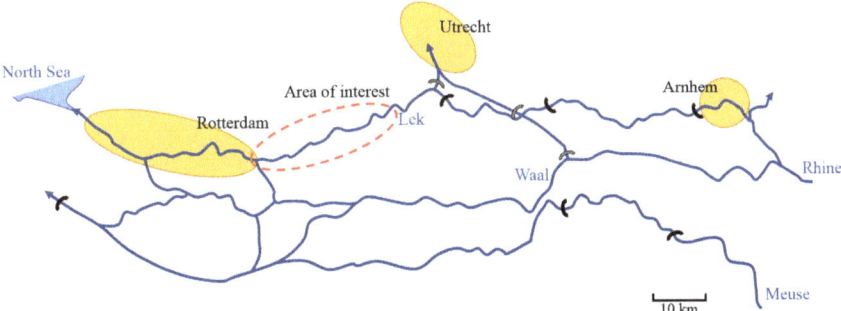

Figure 1. The Rhine–Meuse delta and the location of the area of interest. River flow on the Lek can be controlled by the three weirs in the river. Major cities are depicted by the yellow ovals.

The Rhine–Meuse delta is situated in the western part of the country. Here, the Rhine branches connect with the Meuse river, before flowing into the North Sea through two major outlets. The southern outlet is controlled by sluices (Haringvliet), and the northern outlet is an open shipping channel (Rotterdam Waterway). Sea water enters the estuary through this northern channel.

The western part of the Lek (area of interest; see Figure 1) is a tidal branch of the river Rhine with an open connection to the sea. The discharge is controlled by a weir (Hagestein). During periods of low discharge on the Rhine (below 1500 m^3/s), the Lek receives a minimal net discharge of 1–10 m^3/s. The drinking water inlets under study are situated at two different locations along the downstream section of the Lek, Kinderdijk and Bergambacht (Figure 2, Table 1). Close to Streefkerk, a third location is planned in the near future. At all locations, the type of inlet is river bank filtration: the water is extracted at 40–50 m below surface [6]. Additionally, at Bergambacht, there is an open water intake.

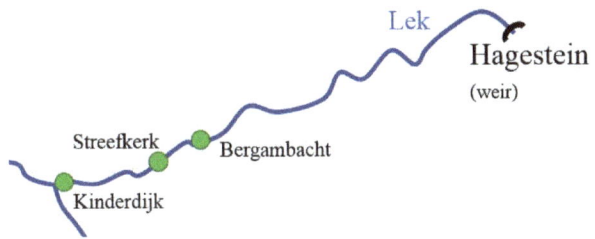

Figure 2. Location of the drinking water inlets along the Lek river.

Table 1. Location and annual extraction volumes of drinking water inlets along the Lek river.

Inlet	Distance to Mouth of Lek	Type of Inlet	Average Annual Extraction
Bergambacht	12 km	(a) Direct (b) River bank filtration	(a) 92 Mm^3 (b) 13 Mm^3
Streefkerk (planned)	8 km	River bank filtration	4–6 Mm^3
Kinderdijk	0.5 km	River bank filtration	6 Mm^3

The mouth of the Lek, just downstream from the intake at Kinderdijk, is situated approximately 42 km from the North Sea. Salt intrusion in the mouth of the Lek commonly occurs during low river flows and high seawater levels. However, it is expected that salinization of the Lek rapidly decays in upstream direction, although few measurement data exist to date to support this view. A typical example of salinization of the mouth of the Lek is presented in Figure 3, showing the situation in the second half of 2018 when a severe hydrological drought occurred in the Rhine river catchment.

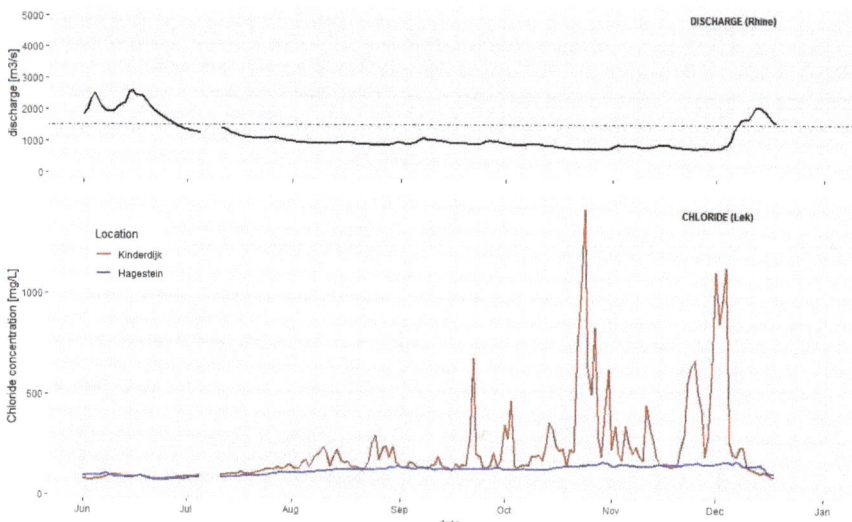

Figure 3. Flow of the Rhine River and salinization of the mouth of the Lek during a hydrological drought in the second half of 2018. Averaged on daily basis but based on 10 min measurements.

Calculations show [7] that the tidal excursion—the distance a water particle travels during a tidal cycle [8]—at the Lek is 6 to 7 km from the mouth under average tidal conditions. This distance increases during spring tide and can reach up to 14 km during storm surges. This implies that the inlet at Kinderdijk is situated within reach of the average tidal excursion. The intake locations at Streefkerk and Bergambacht which are situated more upstream (Figure 2) will only face salinization during storm surges or through mixing processes causing longitudinal dispersion.

The sea is not the only source of salt in the Lek. On the freshwater side, the Rhine carries a salt load as well. The chloride concentration of the Rhine river can be described by the relationship $C(t) = C_b + L_c/Q(t)$, where C_b is the background chloride concentration and L_c is the chloride load. In [9], estimates for C_c and L_c were derived from measurements in 2007–2008 (L_c = 60 kg/s; C_b = 47 mg/L). Using these estimates, the typical chloride concentration for low discharges (800–1500 m^3/s) ranges from 90 to 125 mg/L. A more recent estimate of this riverine chloride concentration (i.e., the combination of background chloride concentration and chloride load), based on the year 2011, results in a range of 97–141 mg/L.

2.2. NDB-Model

For assessing the impact of climate change on salinization of the Lek and the effectiveness of mitigation measures, preferably long time series are calculated in which a large set of variations in conditions like river discharge, tide and wind conditions occur. To date, this can only be carried out with 1D models, which are a commonly applied for hydrodynamic calculations in river studies [10]. Therefore, a 1D hydrodynamic model of the Rhine–Meuse estuary was used to describe the transport of water and salt. This Northern Delta Basin (NDB) model is part of the Dutch National Water Model (NWM), a set of hydraulic and hydrological models and tools set up to support the national fresh water policy [11]. With the NWM model, the hydrology and water distribution throughout The Netherlands can be calculated [12–14]. From this, boundary conditions are extracted for the nested and more detailed NDB model [13–15].

The NDB model is setup in the SOBEK-RE modelling suite, a one-dimensional open-channel dynamic numerical modelling system [16]. Salt transport in the NDB is modelled by a 1D longitudinal advection-dispersion formulation. The advective part describes the distribution of salt along with the 1D motion of the water. Other processes contributing to the distribution of salt that, due to limited dimensions and spatial scale, cannot be resolved by the model are described by the dispersion coefficient. This covers 3D mixing processes like gravitation, circulation and Taylor shear dispersion. Within SOBEK-RE, the dispersion coefficient is estimated by the adjusted version of the Thatcher–Harleman equation [17–19].

The current version of the NDB model (NDB1_1_0) was setup in 2003 [20] and recalibrated in 2005 [21]. Due to its relative long distance from the mouth of the estuary, the river Lek has as yet not been vulnerable to salinization, except for its mouth at Kinderdijk. As a consequence, little data is available and calibration of the NDB model has never focused on the Lek. Only recently, a range for the longitudinal dispersion coefficients was estimated for this part of the Rhine estuary [7], based on an analytic expression for salt dispersion in combination with system knowledge and branch characteristics. An overview of the obtained values is given in Table 2. It shows that the estimate for the dispersion coefficient varies with conditions, like discharge, salinity gradient and location within the estuary. However, the adjusted Thatcher–Harleman formulation in the NDB model is not able to capture this behavior. Therefore, in this study a range of fixed values was used, depending on the minimum upstream discharge at Hagestein (the most right column in Table 2).

Table 2. Overview of the values for the dispersion coefficient K for Kinderdijk and Bergambacht as presented in [7]. All numbers rounded to fives. ΔC is the increase in chloride concentration with respect to the riverine concentration. K_{min} and K_{max} are the minimum and maximum estimate for the dispersion coefficient. Estimates for 40 m^3/s were calculated following the same method as in [7].

Discharge	Location	ΔC = 50 mg Cl/L		ΔC = 500 mg Cl/L		Value Used
m^3/s		K_{min}	K_{max}	K_{min}	K_{max}	
2	Kinderdijk	25	65	30	80	55
	Bergambacht	25	65	30	80	
20	Kinderdijk	30	80	55	125	90
	Bergambacht	25	65	25	70	
40	Kinderdijk	35	90	70	150	110
	Bergambacht	25	65	25	70	

The minimum value used in this study is the average of the estimates for a discharge of 2 m^3/s, i.e., 55 m^2/s. For higher upstream discharges, the dispersion coefficient shows a variation with location along the Lek. The aim of this study is to assess the impact of increasing the upstream discharge on this river branch. To prevent overestimation of the effect of the measure, the dispersion values used were based on the average estimates for the most downstream location (Kinderdijk) and the highest gradient in chloride concentration (ΔC = 500 mg Cl/L).

To assess the impact of this approach, sensitivity calculations have been carried out for an 8-year period. For the reference case, the range of D = 25–80 m^2/s has been explored, which coincides with the full range estimated in Table 2. For the case with a minimum discharge of 20 m^3/s, the range has been extended from 90 m^2/s towards the lowest value estimated in Table 2 (D = 25 m^2/s), since D = 90 m^2/s is expected to be a conservative estimate for the dispersion coefficient, based on typical hydrodynamic and salinity gradient conditions at Kinderdijk. In practice, the dispersion coefficient further upstream of Kinderdijk will be lower.

The results of this sensitivity analysis are shown in Figure 4, in which the 365-day moving average of chloride at Streefkerk is given for the minimum and the 20 m^3/s discharge cases. It shows that the range of the dispersion coefficient is relevant to the results, but that the effect of the upstream discharge is larger, provided the difference in upstream discharge is sufficiently large (some tens of cubic meter per seconds).

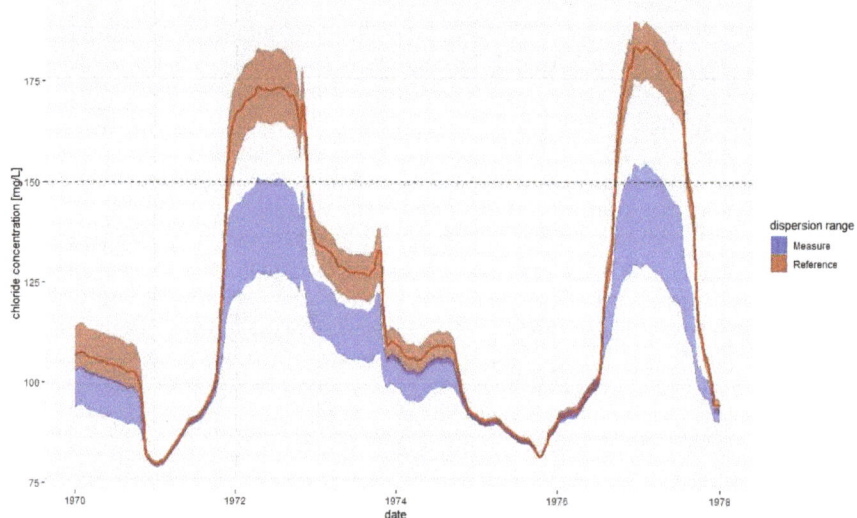

Figure 4. The sensitivity of the 365-day moving average chloride concentration at Streefkerk for the range of the dispersion coefficient: $Q_{up,\,min} = 2$ m^3/s (reference): $D = 25$–80 m^2/s (orange); $Q_{up,\,min} = 20$ m^3/s: $D = 25$–90 m^2/s (blue).

The model setup was validated against observed chloride concentrations at Kinderdijk, available for the period 2001–2011 (no observations were available for the other two locations). Figure 5 illustrates the behavior of the model; it describes the overall variations reasonably well. The model is able to reproduce sudden salinization events due to sea water intrusion. However, the magnitude of the peaks is underestimated.

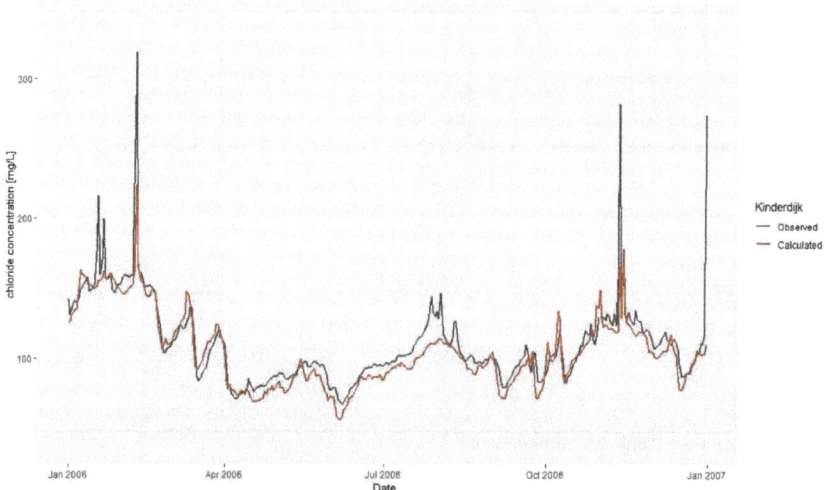

Figure 5. Observed and calculated chloride concentrations at Kinderdijk for the year 2006. Both observations and calculations are daily averaged based on 10 min data. Please note that for the validation runs, measured river chloride concentrations are used, while for the scenario runs (current and future climate), a discharge–salinity relation is used.

This general model performance can also be observed from Figure 6, where the 365-day moving average is plotted for the observed and modelled chloride concentrations at Kinderdijk. The averaged chloride concentration is underestimated for years with a substantial impact of seawater intrusion, like the year 2003.

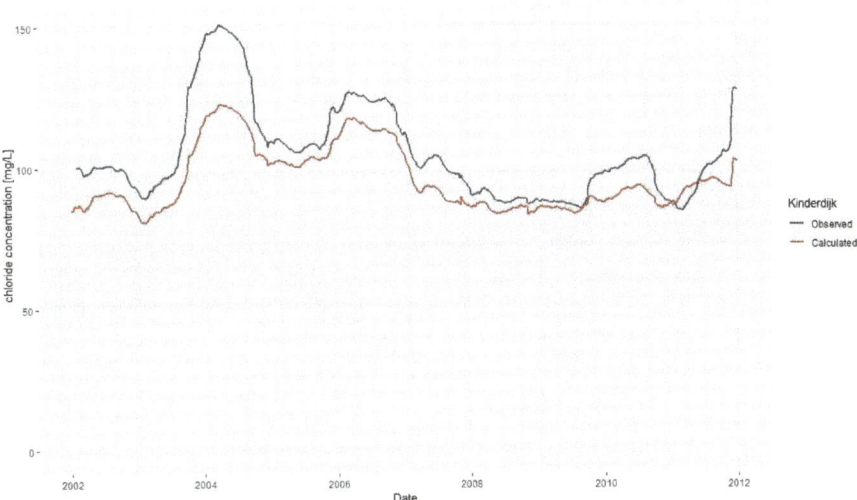

Figure 6. Comparison of the observed and modelled 365-day moving average chloride concentration at Kinderdijk. Please note that the moving average is tailing, that is, the salinization events occurring in the summer and autumn of 2003 start to have a noticeable effect on the MA in the second half of 2003 and remain visible until the second half of 2004.

From this validation, and in line with previous findings [22], it can be concluded that the model is well able to capture salinization events, but that exact variations differ and the influence of sea water intrusion is underestimated. As variations in chloride concentration in the Lek vary between about 50 mg Cl/L up to over 3500 mg Cl/L, estimating exact exceedance durations of a threshold of 150 mg Cl/L requires a very high accuracy of the model. The validation shows that this accuracy cannot be achieved with this basin wide 1D model. In addition, the limited representation of the physical processes relevant for salinity intrusion in 1D poses an uncertainty on the predictability with changing conditions such as sea level rise. However, a global indication on the amount and duration of exceedance in current and future climate can be obtained. The model can therefore be used to carry out a first-order assessment of the vulnerability of the inlet locations to salinization and of the risk reduction that can be achieved by reallocating the available water over the Rhine branches. However, it should not be used in an operational water management context, where more precise estimates are required for a day to day balancing of the freshwater allocation to the Lek and the salinization potential of the intake locations.

2.3. Climate Projection

The climate projection used in this study is the Wh-dry scenario for the Rhine river catchment [1,2]. This scenario is part of the KNMI'14 climate scenarios [23]—a regionalized interpretation of the AR5 climate projections—and serves as the worst case scenario from a fresh water supply perspective. The Wh-dry scenario projects a change in meteorological conditions (precipitation and evapotranspiration) in The Netherlands, impacting the intake and outlet discharges from the river Lek. Furthermore, the Wh-dry scenario projects for 2050 a sea level rise of 40 cm relative to 1995. The Wh-dry scenario projects a strong reduction in summer precipitation in the Rhine catchment by 17% in 2050 [1] and leads to a longer duration and severity of low Rhine river discharges entering The Netherlands. For example, the long term mean annual lowest seven-day flow drops from 1010 m^3/s in current conditions to 825 m^3/s in 2050 in Wh-dry conditions, and the number of days with a flow below 1000 m^3/s doubles from 23 to 46 [2].

To assess the potential impact of climate change under the Wh-dry scenario on salt water intrusion, the NDB model was rerun with adjusted boundary conditions according to the Wh-dry scenario. The 50-year time series of future river discharges and lateral discharges and intakes has been taken from the National Water Model as used in the context of the Delta Program fresh water supply. The projected sea level rise of 40 cm by 2050 has been added to the marine boundary condition of the model thereby copying the variability of tides and storm surges as historically occurred over the 1961–2011 period.

3. Results

3.1. Vulnerability of Drinking Water Inlets

The vulnerability of the drinking water inlets to salt intrusion is indicated by exceedance of the maximum allowable chloride concentration in drinking water in The Netherlands (150 mg/L). For direct inlets, no water is extracted when the concentration of 150 mg Cl/L is exceeded. For the sub-surface inlets (river bank filtration), this maximum allowable concentration is a yearly average. In this section, the vulnerability of the drinking water inlets is analysed in three steps. Firstly, an indication of the increase in salinization on the Lek due to climate change is obtained by analyzing the percentage of time in which the limit of 150 mg Cl/L is exceeded for all three locations along the Lek, for current and future climate. Next, the impact on the 365-day moving average is presented. Finally, the duration of the exceedances is analysed for the direct inlet at Bergambacht.

Figure 7 summarises the number of days that the chloride concentration exceeds the threshold during one or more timesteps in the 50-year period. This exceedance can be caused by either the chloride concentration of the river water (i.e., >150 mg/L) or by seawater intrusion. From these occurrences, the number of days during which the riverine chloride concentration exceeds the threshold are separated. Finally, the occurrences are divided by the total number days in the 50-year period, yielding a percentage of time.

This analysis shows that for reference conditions, the percentage of time at which the threshold chloride concentration is exceeded in the 50-year period considered is close to 2 percent near the mouth (Kinderdijk) and decreases by a factor of 4–5 at the inlet locations upstream. Exceedances of the 150 mg/L threshold due to high chloride concentrations in the river water do not occur. In a future dry climate (Wh-dry conditions), exceedance percentages increase up to 6 to 10 percent, depending on the location. Part of this increase is caused by the riverine chloride concentration, which increases during the low river discharges in the Wh-dry scenario. Please note that this effect alone already causes a 2% exceedance of the threshold at all inlet locations. This is larger than the total exceedance, from marine and river origin, in current climate conditions.

Figure 8 shows the 365-day moving average of the chloride concentration at the three drinking water inlet locations. Under current climate conditions, the moving average does not exceed the threshold of 150 mg/L at any location during the 50 years calculation period. In the Wh-dry scenario, the increase in riverine chloride concentration during low discharges has an effect on the 365-day moving average, but does not lead to exceedances. However, in combination with the increased seawater intrusion, several periods of exceedance at all intake locations occur. In accordance with Figure 7, the number and the duration of exceedances decrease in upstream direction.

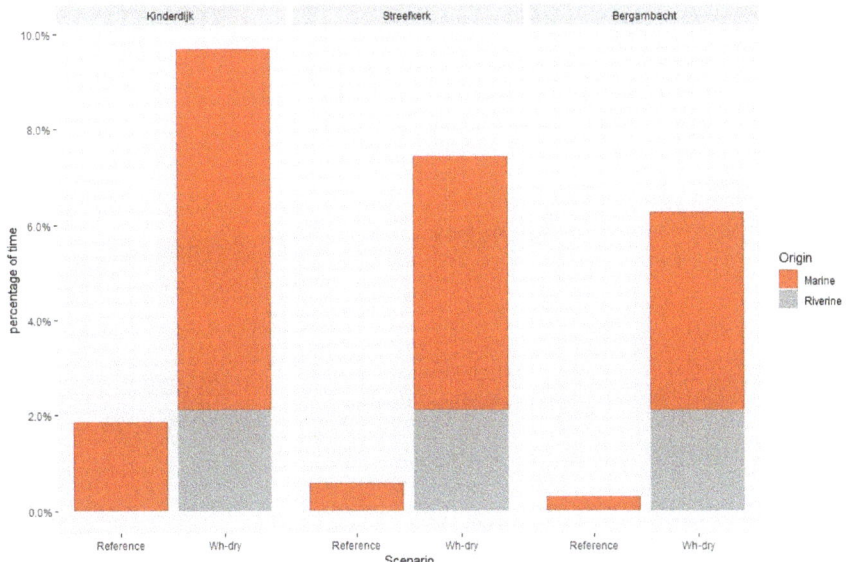

Figure 7. Exceedance of the maximum allowable chloride concentration at the three drinking water inlet locations, based on model calculations 1961–2011, under current conditions (Reference) and Wh-dry conditions. The threshold level is 150 mg/L. Exceedance of the threshold level can be due to a high riverine concentration in the river water (>150 mg/L) or to enhanced seawater intrusion (marine).

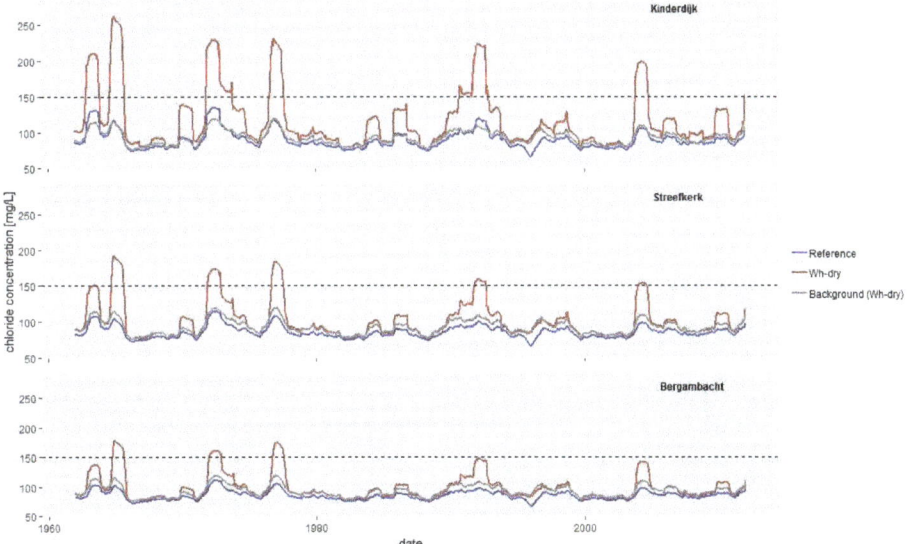

Figure 8. The 365-day moving average chloride concentration at the three drinking water inlet locations (based on calculations 1961–2011). The threshold level is 150 mg/L, averaged over the year, as a moving average.

At Bergambacht, the direct intake of river water is suspended when the chloride concentration exceeds the threshold level of 150 mg/L. This period of suspension can be continued for approx. 20 consecutive days without causing disturbances in the drinking water supply, due to the presence of a freshwater buffer in the coastal dunes. During the 50-year analysis period, in the current climate, only one exceedance with a duration of 20 days or more was identified. In the Wh-dry scenario, this number rises to 17. Table 3 shows the ten longest periods of exceedance. Please note that all salinization events occur in the second half of the year, most of them in the fourth quarter. This is due to the seasonal dynamics of the Rhine river with a minimum flow in the fourth quarter, and to the start of the storm season in the month September, leading to elevated seawater levels at the coast.

Table 3. The 10 longest periods of exceedance of the chloride threshold concentration (150 mg/L) at Bergambacht (direct intake of river water) in the reference situation (current climate) and the future climate Wh-dry scenario. Sorted on highest to lowest durations under Wh-dry conditions.

Nr.	Year	Duration (days) Ref	Duration (days) Wh-dry	Time of Year
1	1976	-	152	Jul–Dec
2	1964	-	116	Jul–Nov
3	2003	1	110	Aug–Dec
4	1971	27	97	Sep–Dec
5	1962	10	77	Okt–Dec
6	1991	1	69	Sep–Nov
7	1990	-	65	Sep–Nov
8	2009	-	54	Sep–Nov
9	1972	7	43	Oct–Nov
10	1985	-	41	Oct–Nov

3.2. Mitigation of Salinization through Adjusted River Water Allocation

The eventual aim of this study is to assess the effects of passing a minimum flow of water through the upstream Hagestein weir (see Figure 2) on the salinization of the drinking water inlets along the Lek. Two variants of this strategy are analyzed in this section: maintaining a minimum discharge of 20 m^3/s and 40 m^3/s, respectively, at Hagestein. This extra water is extracted from the Waal branch, in order to respect the water balance. All other boundary conditions and model settings are unchanged compared to the Wh-dry scenario presented earlier, except for the dispersion coefficient, as described in the method section. The analysis follows the same steps as the previous section.

Analogous to Figure 7, Figure 9 shows the effects of maintaining a minimum discharge of 20 and 40 m^3/s on the Lek river on the number of days in which the chloride threshold of 150 mg/L is exceeded in the Wh-dry scenario (in the 50-year period considered).

The results summarized in Figure 9 show that maintaining a minimum upstream discharge of 20 m^3/s reduces the exceedance time about 1 to 2 percent at the locations Bergambacht and Streefkerk (a 20–25 percent decrease). A minimum discharge of 40 m^3/s decreases this percentage by 30–35 percent. At Kinderdijk, the calculated effect is very small. This is explained by the fact that Kinderdijk is within the normal tidal range of the Lek. An upstream discharge up to 40 m^3/s is very small compared with the volumes of water exchanged during a tidal cycle. Further upstream, the dominance of this alternating advective salt transport diminishes, and the upstream-directed dispersive flux becomes increasingly important for the net longitudinal salt transport. A sufficient increase in the upstream discharge may counterbalance this dispersive salt transport.

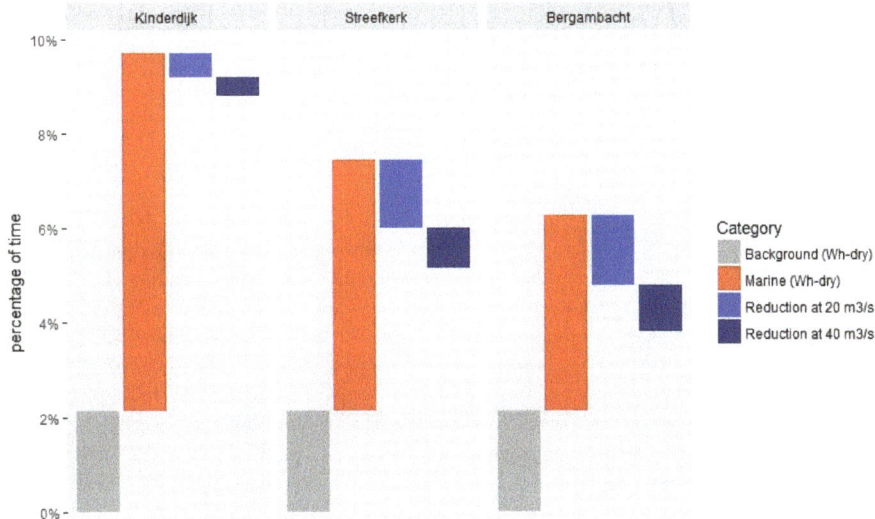

Figure 9. Percentage of time of exceedance of the threshold chloride concentration in the Wh-dry scenario (grey bar plus brown bar) and reduction of this percentage due to maintaining a minimum Lek discharge of 20 and 40 m^3/s (light blue and dark blue bars, respectively).

Figure 10 shows that the annual average chloride concentration at Streefkerk can be kept below the maximum allowable level of 150 mg/L by maintaining a minimum upstream discharge of 40 m^3/s. A minimum discharge of 20 m^3/s also causes a major decrease, but still results in three periods of limited exceedance in the 50 year period.

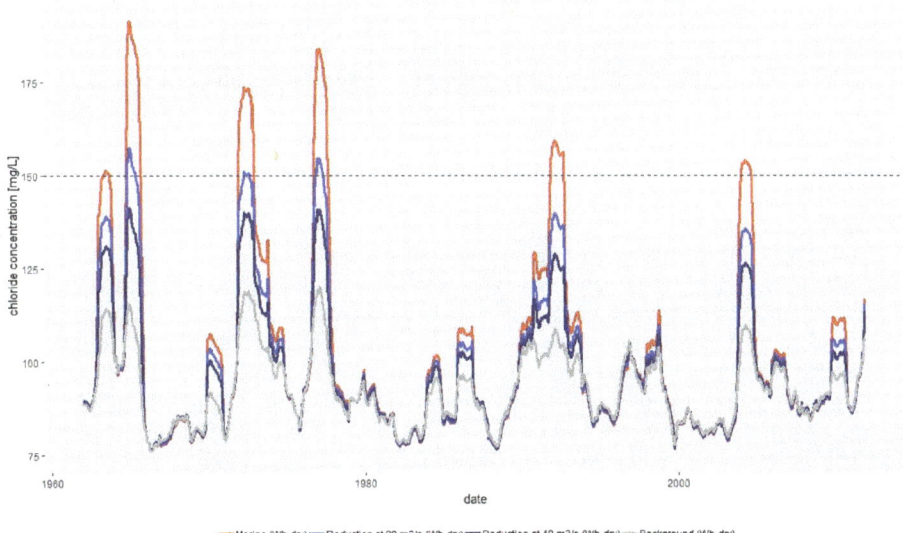

Figure 10. Effect of maintaining a minimum upstream discharge of 20 and 40 m^3/s on the 365-day moving average chloride concentration at Streefkerk 1961–2011.

For the direct inlet at Bergambacht, maintaining a minimum upstream discharge affects both the duration of an exceedance and the maximum chloride concentration. Furthermore, long events can be split into two or more shorter events, as the chloride concentration temporarily drops below the maximum allowable level during a Wh-dry event. Due to this effect, the number of events with a duration of 20 days or more decreases from 17 (with a total length of 1006 days) to 15 (719 days) in the 20 m^3/s variant and 12 (512 days) in the 40 m^3/s variant.

4. Discussion and Conclusions

Below, the results are discussed in view of the uncertainties of the 1D modelling approach and conclusions are drawn. Firstly, the effect of climate change on the salinity intrusion of the Lek and the impact on the drinking water inlets is discussed. Thereafter the effectiveness of the measure is evaluated, including a notion on strategic choices in water distribution to serve different purposes.

Calculations of a 50 year time series show that in the current climate, the instant exceedance of the limit of 150 mg Cl/L is limited (less than ~2% for all locations), the 365-day moving average is at least about 15 mg Cl/L away from exceeding the 150 mg Cl/L and that there is only one event of 20 consecutive days of exceedance of 150 mg Cl/L for the direct inlet at Bergambacht. In the Wh-dry scenario, these numbers show a significant increase. Instant exceedances of the 150 mg Cl/L norm increase from less than a few percent of the time to ~6–10%, the 365-day moving average exceeds the 150 mg Cl/L threshold for several times and at the direct inlet at Bergambacht over 10 periods of long-term exceedance of the limit are calculated. Though the validation showed that exact exceedance numbers and duration of the threshold of 150 mg Cl/L cannot be obtained from the 1D model, a global indication can be retrieved. It can therefore be concluded that in a current climate, the water inlets rarely face problems and that this will change in future climate. As there are indications that the chosen 1D approach may lead to a relatively low response to mean sea level rise [24], the influence of climate change on the salt intrusion and related exceedance times may even be stronger.

Another important finding is that salinization on the Lek is not solely related to the connection to the sea. The chloride load on the Rhine has been strongly reduced since the 1960s [19], such that in current climate, even during low discharge events, when the chloride load is less diluted, the maximum allowable concentration for drinking water is hardly ever exceeded (Figure 4). In the Wh-dry scenario, the chloride concentration of the Rhine increases considerably during low river flows, accounting for about 2% of the exceedance. In contrast to the influence of chloride of marine origin, which mostly affects the downstream locations, riverine chloride affects all stations equally. It should be noted that the relationship between discharge and chloride concentration at the German-Dutch border used in this study [13] is based on measurements from 2011. As between 1997–2008 a rapid decrease was observed in the chloride load on the Rhine [9] it is relevant to know whether further changes have occurred in recent years.

Calculations show that maintaining a minimum upstream discharge of several tens of cubic meters per second reduce the risk of salinization at the inlet locations. However, the increased influence of seawater intrusion on the drinking water inlets cannot be fully compensated for by this measure. The increased upstream discharge is most effective in counteracting the inward salt transport by mixing processes that cause longitudinal dispersion, but less effective in counteracting the salt transport caused by tides and storm surges, as the extra discharge is small compared to the large volumes of water exchanged during these events. Consequently, the effect at Kinderdijk is limited while further upstream at Streefkerk and Bergambacht, several tens of cubic meters per second on the Lek reduce the salinity intrusion events to the level that will cause limited hindrance to the fresh water intake in the Wh-dry scenario.

This study aims at assessing the risk of exposure of the drinking water inlet locations to salinization and of the reduction that can be achieved by reallocating the available water over de Rhine branches. It can be concluded that diverting water onto the Lek is an effective measure to reduce the risk of salinization at Bergambacht and Streefkerk. Kinderdijk is well within the tidal excursion of the Lek and cannot profit from a relatively small upstream surplus. For individual events however, the operational question of how much water should be passed through the weir at Hagestein cannot be answered by the results presented here. This requires more precise estimates, which are expected to be obtained by carrying out a hybrid 1D and 3D approach. Given the more detailed representation of the physical processes within a 3D model, time slices of the 1D result could be selected and recalculated. By this means, estimates for the effect of climate change on the salinity intrusion during particular low river-discharge events and the required discharge surplus can be improved.

The extra water discharged onto the Lek comes at a cost. It cannot be used elsewhere in the river and adjacent channel system to sustain fresh water demands for water quality (flushing), irrigation and navigability. Also, the extra water is extracted from the Waal river, which is the main inland shipping channel for the port of Rotterdam. During low discharges of the Rhine, water levels on the Waal are very critical, as they determine the allowed depths of ships and hence the loads they can carry. The potential gain of the extra water for the drinking water inlets along the Lek has to be balanced against the impact of this measure on water level and stream flows in other parts and functions of the delta system. This calls for more precise estimates of both the climate effect and the amount of discharge needed for particular events to counteract the salinity intrusion.

Author Contributions: Conceptualization, M.B. and G.Z.; Methodology, M.V., Y.H. and M.B.; Supervision, M.B.; Validation, Y.H.; Visualization, M.V. and G.Z.; Writing—original draft, M.V. and Y.H.; Writing—review & editing, M.B. and G.Z.

Funding: This study was funded by the National Association of Water Companies in The Netherlands (VEWIN) and the water companies Oasen and Dunea.

Acknowledgments: The authors would like to thank A. Kersbergen at HydroLogic for her diligent work in processing large amounts of data into useful insights. Also, we would like to thank M. Mens at Deltares and N. Kielen at Rijkswaterstaat for their valuable comments during the preparation of this article. We further thank C. Kuyper for his valuable input on the system dynamics of salinity intrusion on the Lek.

Conflicts of Interest: One of the co-authors (Gertjan Zwolsman) is employed at water company Dunea, which financed 25% of this study. He contributed to the design of the study and the visualization of the results, but was not involved in the data collection and modelling part of the study. In addition, he reviewed the MS on style and consistency, but had no influence on the interpretation of the model results.

References

1. Lenderink, G.; Beersma, J. *The KNMI'14 WH, Dry Scenario for the Rhine and Meuse Basins*; KNMI: De Bilt, The Netherlands, 2015.
2. Sperna Weiland, F.; Hegnauer, M.; Bouaziz, L.; Beersma, J. *Implications of the KNMI'14 Climate Scenarios for the Discharge of the Rhine and Meuse, Comparison with Earlier Scenario Studies*; Deltares: Delft, The Netherlands, 2015.
3. Klijn, F.; Ter Maat, J.; Van Velzen, E. *Zoetwatervoorziening in Nederland, Landelijke Analyse Knelpunten in de 21e Eeuw*; Deltares: Delft, The Netherlands, 2011. (In Dutch)
4. Mens, M.; Van der Wijk, R.; Kramer, N.; Hunink, J.; de Jong, B.J.; Becker, P.; Gijsbers, P.; Ten Velden, C. *Hotspotanalyses voor Het Deltaprogramma Zoetwater: Inhoudelijke Rapportage*; Deltares: Delft, The Netherlands, 2018; p. 149. (In Dutch)
5. Van Waterstaat, M.I. En Delta Programme—Government.nl. Available online: https://www.government.nl/topics/delta-programme (accessed on 19 February 2019).
6. Van der Kooij, D. *Drinkwater uit Oevergrondwater: Hydrologie, Kwaliteit en Zuivering*; Mededeling; Keuringsinstituut voor Waterleidingartikelen: Nieuwegein, The Netherlands, 1985. (In Dutch)
7. Kuijper, C. *Analyse van Zoutmetingen in de Lek, Met Schatting Dispersiecoefficient*; Deltares: Delft, The Netherlands, 2017; p. 76. (In Dutch)
8. Savenije, H.H.G. *Salinity and Tides in Alluvial Estuaries*, 2nd completely revised ed.; version 2.5; Delft University of Technology: Delft, The Netherlands, 2012.
9. Bonte, M.; Zwolsman, G. Klimaatverandering en verzoeting van de Rijn. *H2O* **2009**, *3*, 29–31. (In Dutch)
10. Cunge, J.A.; Holly, F.M.; Verwey, A. *Practical Aspects of Computational River Hydraulics*; Pitman: Boston, MA, USA; London, UK; Melbourne, Australia, 1980; ISBN 978-0-273-08442-6.

11. Rijkswaterstaat National Water Model. Available online: https://www.helpdeskwater.nl/onderwerpen/applicaties-modellen/applicaties-per/watermanagement/watermanagement/nationaal-water/technische/zoetwaterverdeling/ (accessed on 7 November 2018).
12. Prinsen, G.; Sperna Weiland, F.; Ruijgh, E. The Delta Model for Fresh Water Policy Analysis in The Netherlands. *Water Resour. Manag.* **2015**, *29*, 645–661. [CrossRef]
13. Hunink, J.; Hegnauer, M. *Update Deltascenario's Nationaal Water Model*; Deltares: Delft, The Netherlands, 2015. (In Dutch)
14. Hunink, J.; Delsman, J.; Prinsen, G.; Bos-Burgering, L.; Mulder, N.; Visser, M. *Vertaling van Deltascenario's 2017 Naar Modelinvoer voor Het Nationaal Water Model*; Deltares: Delft, The Netherlands, 2018. (In Dutch)
15. Deltares Nationaal Water Model—Nationaal Water Model—Deltares Public Wiki. Available online: https://publicwiki.deltares.nl/display/NW/ (accessed on 27 February 2019).
16. *Salt Intrusion, Technical Reference*; Sobek-RE Documentation; Deltares: Delft, The Netherlands, 2012.
17. Thatcher, M.; Harleman, D. A mathematical model for the prediction of unsteady salinity intrusion in estuaries. In *R.M. Parsons Laboratory Report*; MIT: Cambridge, MA, USA, 1972.
18. Rijkswaterstaat. *IJking Chloridedeel ZWENDL Noordelijk Deltabekken (Stand van Zaken September 1984)*; Rijkswaterstaat: Rotterdam, The Netherlands, 1984. (In Dutch)
19. Huismans, Y.; Buschman, F.; Wesselius, C.; Daniels, J.; Kuijper, C. *Modelleren van Zoutverspreiding Met SOBEK 3 en SOBEK-RE*; Deltares: Delft, The Netherlands, 2016. (In Dutch)
20. Kraaijeveld, M. *Een SOBEK-Model van Het Noordelijk Deltabekken; Kalibratie en Verificatie Zoutbeweging Noordrand*; Rijkswaterstaat: Dordrecht, The Netherlands, 2003. (In Dutch)
21. Jansen, M.H.P.; Collard, E.A. *Herkalibratie van de Zoutverdeling NDB-Model Fase 2*; Svasek Hydraulics: Rotterdam, The Netherlands, 2005. (In Dutch)
22. Huismans, Y.; van der Wijk, R.; Fujisaki, A.; Sloff, C.J. *Zoutindringing in de Rijn-Maasmonding: Knelpunten en Effectiviteit Stuurknoppen*; Deltares: Delft, The Netherlands, 2018. (In Dutch)
23. KNMI. KNMI Climate Scenarios. Available online: http://www.climatescenarios.nl/scenarios_summary/index.html (accessed on 7 November 2018).
24. Daniels, J. Dispersion and Dynamically One-Dimensional Modeling of Salt Transport in Estuaries. Master's Thesis, Delft University of Technology, Delft, The Netherlands, National University of Singapore, Singapore, 2016.

© 2019 by the authors. Licensee MDPI, Basel, Switzerland. This article is an open access article distributed under the terms and conditions of the Creative Commons Attribution (CC BY) license (http://creativecommons.org/licenses/by/4.0/).

Communication

Climate Change and Extreme Weather Drive the Declines of Saline Lakes: A Showcase of the Great Salt Lake

Qingmin Meng

Department of Geosciences, Mississippi State University, Starkville, MS 39762, USA; qmeng@geosci.msstate.edu

Received: 5 December 2018; Accepted: 22 January 2019; Published: 23 January 2019

Abstract: A viewpoint of a temporal trend with an extremely changing point analysis is proposed to analyze and characterize the so-called current declines of the world's saline lakes. A temporal trend of a hydrological or climate variable is statistically tested by regressing it against time; if the regression is statistically significant, an ascending or declining trend exists. The extremely changing points can be found out by using the mean of a variable, adding or subtracting two times of its standard deviation (SD) for extremely high values and extremely low values, respectively. Applying the temporal trend method to the Great Salt Lake's (GSL) relationship between its surface levels and precipitation/temperature in the last century, we conclude that climate changes, especially local warming and extreme weather including both precipitation and temperature, drive the dynamics (increases and declines) of the GSL surface levels.

Keywords: dynamics of saline lakes; extremely changing points; extreme weather; temporal trend

1. Introduction

The declines of saline lakes were recently highlighted in research and media. The Great Salt Lake (GSL), a remnant of Lake Bonneville, existed from about 30,000 years ago to 16,000 years ago, and is now approximately 4402.98 km^2 (1700 square miles) with a length of 120.70 km (75 miles) and a width of 45.06 km (28 miles) at its average water level [1]. It has no outlet, with dissolved salts accumulated by evaporation. Laying on a shallow playa, small changes in water surface levels typically result in large changes of the GSL area. The lake drainage basin is about 90,649.58 km^2 (35,000 square miles), where the human population is now more than 1.5 million. The GSL seems to be an ideal lake to study to understand the impacts of changes in climate on water resources.

Human water use might be an important factor driving the declines of world saline lakes. Using the GSL as an example, some researchers concluded that human water uses, specifically consumptive water uses for agricultural, salt pond mineral production, and municipal and industrial purposes determine the declines of saline lakes [2]. Although the US freshwater withdrawals have declined since 1980, (i.e., Trends in estimated water use in the United States, 1950–2015, https://water.usgs.gov/watuse/wutrends.html), and the current consumptive water uses in agriculture, salt pond mineral production, and industry can be much larger than that in the 1950s, in the last century the GSL had experienced a number of times of significant continuous declines, such as 1925–1936 and 1952–1963; furthermore, the current 2016 decline is much better than its situation in the 1960s. In other words, in the long term, human water use could be important, but it is questionable to only just attribute saline lakes' decline to human/consumptive water uses (i.e., agricultural, mineral municipal aspects, and others).

The Landsat images, which are available from 1972 to current years, as shown in Figure 1, indicate that the worst decline situation was in 1972 compared to 1987, 1999, 2011, and 2016. Human water use, including agricultural, salt pond mineral, and municipal and industrial purposes in the 1970s was much less than those in the 1980s, 1990s, and 2010s. Human water use, thus, is not the main driving force of declines in the GSL in the past 100 years.

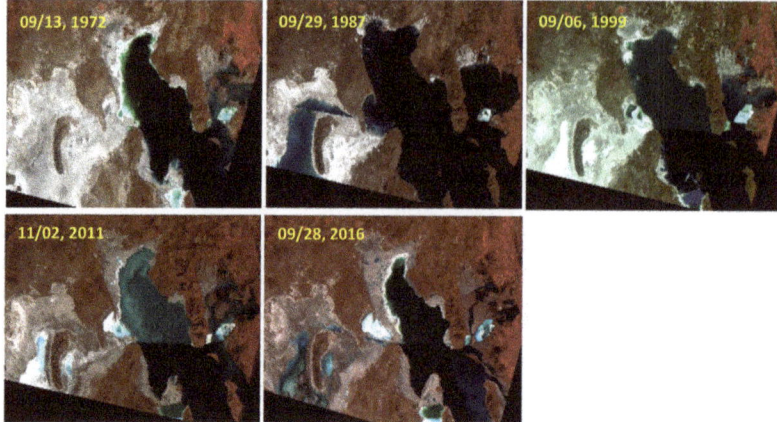

Figure 1. Landsat images (1972, 1987, 1999, 2011, and 2016) displayed in false color for the Great Salt Lake (GSL) and the GSL desert that is on the left side of the lake in the imagery. Sources and more details are available at http://Earthshots.usgs.gov, Earthshots: Great Salt Lake.

The recent changes of GSL water levels in the last three years 2016, 2017, and 2018 (details are available at http://greatsalt.uslakes.info/Level.asp) further reject the above conclusion that human water use resulted in the GSL water loss. The average water levels in 2017 or 2018 are 0.91 meters higher than those in 2016. Given the current lake area 4402.98 km^2, it means that about 4,006,711,800 m^3 more water was added into the GSL in 2017 or 2018 than in 2016. Given the fact that the changes of human water use in 2016, 2017, and 2018 are barely due to no significant changes in population, agriculture, industries, and other human activities, human water use, hence, is not the dominant factor for GSL water loss or water level dynamics in the short term.

Considering the water budget of the GSL, we define the GSL water level using the equation below: GSL water level = inflow (precipitation + river discharge)—outflow (human water use + evaporation). As discussed above, human water use alone cannot be thought of as the main driving force for water level dynamics or water loss. Thus, there are three remaining factors of precipitation, river discharge, and evaporation, which are all mainly related to climate factors of precipitation and temperature. Given this information, it is necessary to rethink the impacts of climate on the dynamics or declines of GSL water levels. Precipitation is the main and direct water source, and evaporation caused by the increases in temperature can be the dominant water loss of saline lakes.

2. Extremely Changing Point Analysis and Data

This short research proposes a new viewpoint of a temporal trend with an extremely changing point analysis in order to analyze and reveal the so-called declines of saline lakes. To the best of our knowledge, there is not a conception of an "extremely changing point analysis" in the literature, and hence it is proposed for applications to hydrology, climatology/meteorology, and environmental study to efficiently identify significantly changing observations in a data series.

In both climatology/meteorology and hydrology, time series are the data that are often analyzed, and extremely large or small records typically have their specific meaning in climate or hydrology. For instance, extreme weather has recently received more attention than before [3], and in the USA it is defined as unusual or unexpected severe weather at the extremes of the historical distribution, which typically are in the most unusual ten percent [4]. Studies have indicated that extreme weather in the future will pose an increasing threat to the world, and three times the standard deviation were used to indicate extremely hot summer outliers [5]. Extreme values are key aspects of climate change, and changes in extremes are typically the most sensitive climate characteristics for ecosystems and societal responses [3,6]. Extremely increased or decreased records can be significantly large for seemingly modest mean changes in climate [7]. Most climate impacts mainly result from extreme weather events or the climate variables that are significantly above or below some critical levels, which hence affect biological behaviors or the performance of physical systems [3,6,8]. Intergovernmental Panel on Climate Change (IPCC) further stated that for important climate impacts, scientists are interested in the effects of specific extreme events or threshold magnitudes [3]. Therefore, it is necessary to apply an extremely changing point analysis to reveal the declines of GSL.

2.1. Extremely Changing Point

Whether a record is an extremely changing point is identified by using the mean of an attribute adding or subtracting two times of its standard deviation (SD). If an observation is larger than its mean plus two SD, it is an extremely high value point, while an observation smaller than its mean minus two SD is an extremely low value point.

Here, the extremely changing point analysis is based on the common statistical concept of the Z-score (or standard score) in statistics, which is defined as $Z = \frac{x-u}{\sigma}$, or written as $x-u = Z*\sigma$ that is easily explained and understood, where x is the observations of an attribute u is the mean of the population, and σ is the standard deviation of the population; it describes how the observations are off the population mean. When Z-score is related to a normal distribution [9,10], the Z-score ranges from −3 to +3 covering almost the whole distribution by approximately 99.7%. The extremely changing point analysis proposed in this study is not limited by normal distribution, which in reality is often a special case, and for example, both mean and standard deviation require attention in order to understand extreme temperature [11]. Both two times the standard deviation and three times the standard deviation have been used to examine extreme temperature [5], although the gamma distribution is often used to model temperature measurements. In this study, we define $Z = 2$ to determine if a record is far away enough from the mean, so that it can be defined as an extremely changing point; x can be either 2σ larger than the mean or 2σ smaller than the mean, although this method is used in statistics to find outliers.

As we know, a temporal trend can be statistically tested by an attribute regressed against time. If a trend is statistically significant with a p-value of its slope test less than 0.05, a positive slope indicates an increasing trend, while a negative slope indicates a declining trend.

We often emphasize a general trend for an environmental phenomenon, while extremely changing points (such as extremely high or low temperatures) and their effects are often overlooked, but they play significant and increasing impacts on the environment and human society as environmental changes become more global and frequent. Extremely changing point analysis not only adds more properties characterizing an environmental phenomenon that could not be disclosed by trend analysis, or mean as well as variance analysis, but also could reveal the relationship between geographic phenomena, such as extremely high precipitation which typically results in significant flooding inundation in space and time.

2.2. The Data

Using the GSL as a case study, the lake surface level data and climate data (including temperature, precipitation, and snowfall) are processed first from 1904 to 2016, given the snowfall data are available from 1904. The mean and SD of the four variables are calculated; then, the lower bound (i.e., mean minus two SD) and upper bound (mean plus two SD) are used as thresholds to respectively determine the extremely low and the extremely high values in each time series data of lake surface level, precipitation, temperature, and snowfall. Based on the lower bound and upper bound, the extremely low points and extremely high points of these four variables are recognized and summarized in Table 1.

Table 1. Climate, extreme weather, and the surface levels of the Great Salt Lake.

	Mean	2* SD	Lower Bound	Upper Bound	The Extremely Low Observations	The Extremely High Observations
Surface Level Meter	1279.82	2.33	1277.49 (1278.23)	1282.15 (1281.41)	1277.76 in 1963	1282.52, 1282.87, 1283.34, 1283.31, 1282.54 in 1984–1988
Precipitation Meter	0.395	0.173	0.223 (0.26)	0.568 (0.54)	0.221 in 1979	0.581, 0.616, 0.605 in 1982, 1983, 1998
Temperature K	284.31	257.15	282.53 (282.88)	286.09 (285.77)	282.15, 259.98 in 1964, 1983	286.76, 286.2, 286.65, 286.59 in 2012, 2014, 2015, 2016
Snowfall Meter	1.399	1.088	0.311 (0.57)	2.488 (2.29)	0.274 in 1973	2.682, 2.578, 2.979, 2.49, 2.51 in 1916, 1921, 1951, 1983, 1992

Note: The lower bound is determined by the mean minus two standard deviations (SDs), and the upper bound is the mean plus two SDs. Surface level data of the GSL are available at https://ut.water.usgs.gov/greatsaltlake/elevations/. Precipitation, temperature, and snowfall data are available at https://www.ncdc.noaa.gov/IPS/lcd/lcd.html. More details about these downloaded data are summarized in the Supplementary Materials. The values in ( ) are the 5th and 95th percentiles, respectively, compared with those lower bounds and upper bounds determined by mean and 2SD.

Some may question, why not use 5th and 95th percentiles to identify extremely changing points? In Table 1, the values of the 5th and 95th percentiles for surface level, precipitation, temperature, and snowfall are compared to the lower and upper bounds respectively defined by 2SD. Results show that the values of 95th percentiles are all much smaller than the upper bounds determined by mean + 2SD, as shown in Table 1, but the 5th percentiles are all much larger than the lower bounds determined by mean—2SD. In other words, if the 95th percentile is used, there could be too many "extremely changing points" that in fact are not large enough to be extreme; while many more low values also can be added by using the 5th percentile, which are not small enough to be extremely low. Therefore, the 5th and 95th percentiles cannot make the extremely changing points as meaningful as it is defined to identify the extremely changing observations by using mean and 2SD. For instance, using the 5th and 95th percentiles, surface levels in 1923, 1924, and 1989 would be added as extremely high water levels to 1984, 1985, 1986, 1987, and 1988; surface levels in 1961, 1962, 1964, 1965, 2015, and 2016 would be added as extremely low observations in addition to 1963, as shown in Figure 2. In other words, mean and 2SD are more robust and effective than the 5th and 95th percentiles to identify extreme values.

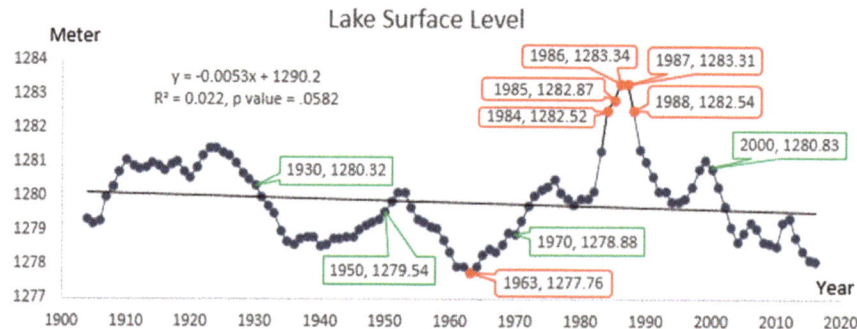

Figure 2. The general declining trend of the GSL surface level and the extremely changing points marked in red.

From the early 1970s, there is a significant trend of local climate warming in the GSL region, which is primarily driving the declines of the GSL, as shown in Figure 3. Therefore, the climate variables of temperature, precipitation, and snowfall are analyzed into two periods, from 1904 to 1970 and 1971 to 2016, in which temporal trends are analyzed and extremely changing points are marked in red, as shown in Figure 3.

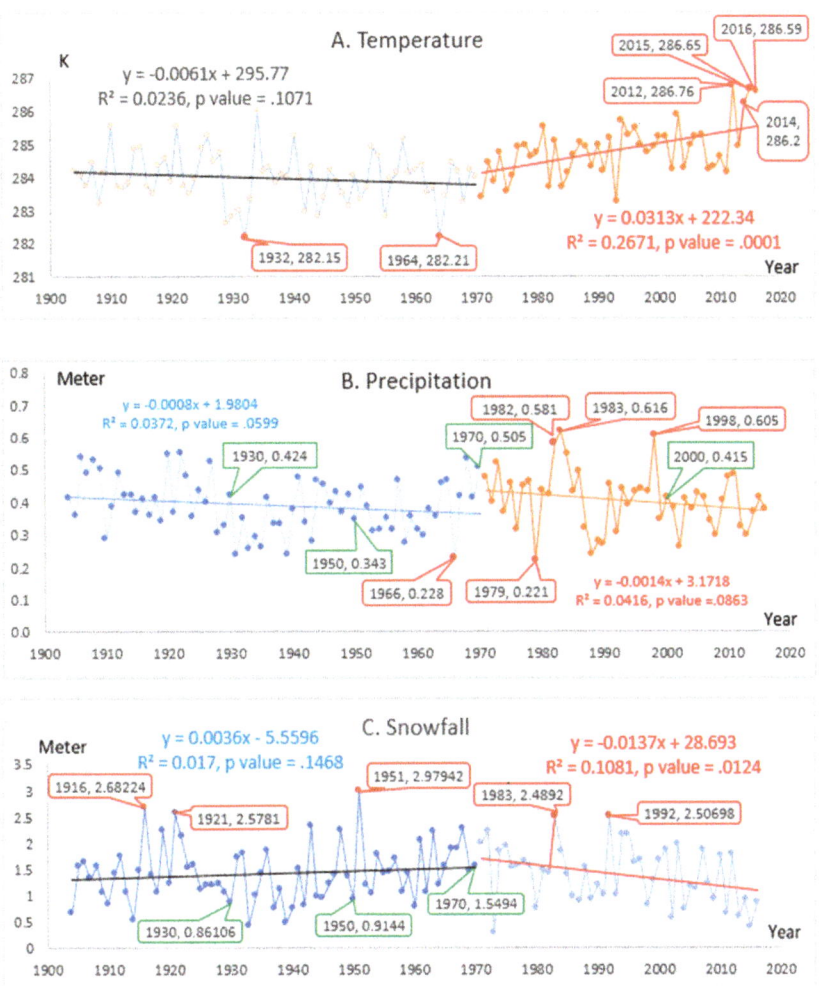

Figure 3. The dynamics of temperature, precipitation, and snowfall in the GSL area. (**A**) trend analysis of temperature; (**B**) trend analysis of precipitation; and (**C**) trend analysis of snowfall.

Trend analysis is also applied to the lake surface level, temperature, precipitation, and snowfall in order to highlight the efficiency of extremely changing point analysis in hydrological and climate data analysis. For example, whether the GSL has a significant declining trend can be analyzed as a regression model with years as a predictor. If we use the lake surface level as an indicator of GSL declines, we need to regress the lake surface levels against years. The results of trend analysis and the extremely changing points are plotted and marked in red, as shown in Figures 2 and 3.

3. Results and Discussions

In the last century, the GSL shows an apparent decreasing trend from 1904 to 2016 with the lowest level of 1277.76 meters in 1963, but its declining trend is not significant, as shown in Figure 2. In other words, the significance of GSL declines is tested with a *p*-value of 0.0582 that is just a little beyond the significance level 0.05. However, the periodic changes of its surface levels in both increasing patterns and decreasing patterns are anomalous in the last century, as shown in Figure 2. This cannot be explained by the human water uses in agriculture, industry, and municipal purposes that were much lower in the last century than current days given the recently significant growth in agriculture, industry, and urbanization in the GSL basin. Instead, the GSL declines could be more clearly explained by climate change and extreme weather including both precipitation and temperature, which then are explained in the five sections below: local warming and evaporation, the impacts of radiation and wind on evaporation, precipitation, river discharge, and human water use.

3.1. Local Warming and Evaporation

Climate changes, especially increasing temperature, have caused significant water loss through evaporation in semi-arid regions [12–16]. Craig et al. reported that increasing evaporation rates caused by climate warming have resulted in approximately 40% of Australia's total water storage capacity loss every year [16]. Helfer et al. and Johnson and Sharma obtained similar results of climate warming on evaporation, demonstrating that increasing temperature results in the significant increases of annual average evaporation [13–15].

Specifically, Helfer et al. mentioned that a temperature increase range between 0.8 °C and 1.3 °C in 2030–2050 compared to 1990–2010 will result in an average annual evaporation from 1300 mm in 1990–2010 to an average annual 1400 mm in 2030–2050 [13]; in other words, an annual temperature increase rate of 0.02 to 0.0325 would result in an average annual difference of 100 mm in evaporation. Additionally, an annual temperature increase rate of 0.021 to 0.04 would cause an average annual evaporation difference of 190 mm, i.e., evaporation from 1300 mm in 1990–2010 to 1490 mm in 2070–2090 caused by a temperature increase range of 1.7 °C to 3.2 °C in 2070–2090 [13]. In the GSL region, the annual temperature increase rate is 0.0313 from 1971 to 2016, as shown in Figure 3A. Given an average annual evaporation amount E for the GSL before 1960 and the significant and consistent temperature increase from 1970 to 2016, as shown in Figure 3A, the average annual evaporation in 1970–1990 could be $E + 100$ mm, and the average annual evaporation in 2000–2020 could be $E + 190$ mm. Therefore, the GSL has lost a huge amount of water from evaporation in the last 50 years. The extremely high temperatures in the recent years 2012, 2014, 2015, and 2016 directly aggravate the significant declines of surface water levels of the GSL due to the enhanced evaporation with the addition of low precipitation in the recent few years, and the GSL hence reaches a relatively low level of 1278.13 meters in 2016, as shown in Figures 2 and 3. Therefore, evaporation and low precipitation are the main cause of the declines of the GSL.

3.2. Radiation, Wind Speed, and Evaporation

Some may still wonder if radiation and wind speed impact evaporation changes. Solar radiation increased significantly in the last century [17], which indicated that evaporation increased too. The effects of wind speed on evaporation are complex. At low velocity values, the first stage evaporation rate will increase when wind speed increases, but at the same time the transition time decreases; however, at high values of wind speed, evaporation rates will depend less on the wind speeds; additionally, no significance is found for the impact of the wind speed on the second stage evaporation rate [18]. In all, the effects caused by increasing radiation could be stronger than the effects of wind speed, and thus evaporation might increase in the last century; or the overall evaporation change caused by radiation and wind speed is not significant in the last century.

3.3. Precipitation

The temporal patterns of precipitation directly drive the general dynamics of GSL surface levels, which are then mainly modified by evaporation driven by temperature changes as discussed above. Both the periods from 1904 to 1970 and from 1971 to 2016 show apparent decreases in precipitation, but the significances are just above the significance level 0.05. Therefore, the declining trend of lake surface level is not quite significant. Since 1971 with about 10 years' high precipitation, the GSL reached an extremely high level in 1984 that continued to 1988, as shown in Figure 2, because of the extremely high precipitation in 1983, 1984, and the high precipitation in 1985 and 1986, as shown in Figure 3. Additionally, from 1904 to 1930, the GSL surface levels were most often above its average value (1279.82 meters), as shown in Table 1, because the precipitation in only 8 of 27 years was relatively lower than its average value of 0.395 meters, as shown in Figure 3. In the 40 years from 1931 to 1970, only two year's GSL surface levels (1952 and 1953) were relatively above its average level, because the precipitation in 27 years was much lower than the average precipitation, as shown in Figure 3; especially with 20 more years of low precipitation, and in 1963, the GSL reached its lowest surface level (1277.76 meters) that was still above its extremely low level bound (1277.49 meters). From 1971 to 2000, there were only 9 of the 30 years when precipitation was less than its average value, and therefore most of the years the GSL surface levels were above its average level. The extremely high precipitation in 1998 resulted in the second highest surface level in 1999, as shown in Figures 2 and 3. Although the precipitation in 1966 and 1989 was very low, respectively high precipitation following them continuously occurred for 7 or 8 years, which thus did not result in extremely low surfaces but relatively low water levels.

The continuously significant decreases in snowfall from 1971 to 2016 could be a secondary contribution besides the significant increasing temperature for the apparent declines of lake surface levels after 2000, as shown in Figures 2 and 3. The extremely high snowfall in 1983 (2.489 m) also secondly contributed to the extremely high records of lake surface levels in 1984 to 1987.

3.4. River Discharge

Bear River, Jordan River, and Weber River are the major surface water discharges into the Great Salt Lake, which account for 58%, 22%, and 15%, respectively, for the total inflows of the lake [2]. The U.S. Geological Survey (https://waterdata.usgs.gov/) provides inflows from 1971 to 2018 for the Bear River and Joran River as displayed in Figure 4 below, which have 80% of surface inflows for the GSL. Because the continuous data for Weber River are only available after 1989, it is not included into this decadal analysis. In general, the surface inflows show a significant declining trend, which contributes to the decreasing water level in the same period, as shown in Figure 2. The reduced river discharge is directly caused by the declining precipitation and snowfall, as shown in Figure 3.

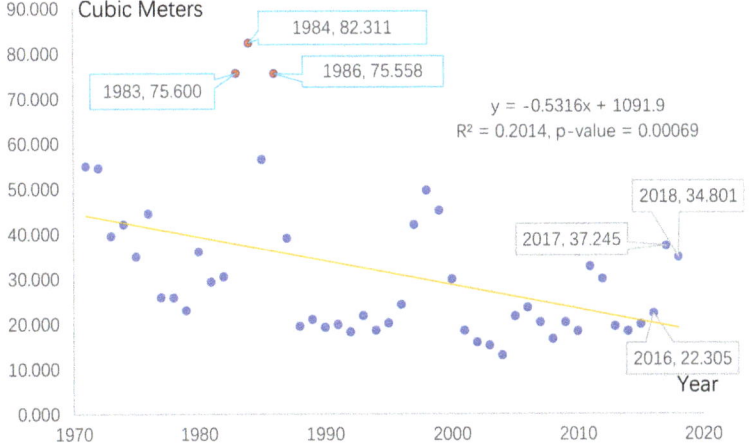

Figure 4. Surface inflows from Bear River and Jordan River into Great Salt Lake from 1971 to 2018. Data source: https://waterdata.usgs.gov/.

The extreme highest points of the observed discharge values in 1983, 1984, 1986, as shown in Figure 4, have significant contribution to the extremely high water levels, as shown in Figure 2. These extreme highest values are also coincident to the extremely high precipitation values observed in 1982 and 1983. This is a truth that we first observed highest precipitation, secondly observed highest surface inflows, and third detected highest lake water levels. In the GSL basin, climate changes generally drive the river discharge patterns.

3.5. Human Water Use

We understand that human water use could be another secondary contribution to GSL's water loss. However, the conclusion of "consumptive water use including agricultural, salt pond mineral production, and municipal and industrial uses rather than long-term climate change has greatly reduced its size for the Great Salt Lake's surface declines" [2], cannot explain the much lower surface levels in 1961, 1962, 1963, and 1964 that were lower than 2016's current record, the significant declines from the early 1920s to the late 1930s, the significant increasing trend from the mid-1960s to the mid-1970s, and other significant declining or increasing patterns in the last century, as shown in Figure 2.

Additionally, human water use even cannot explain the current GSL water level changes in 2016, 2017, and 2018, as shown in Figure 5. The average water levels in 2017 or 2018 are 0.91 meters higher than that in 2016. If human water use was the dominant driving force for GSL water loss or dynamics, it is impossible that in three continuous years, significantly less human water use has occurred, which then has resulted in such a huge amount of water increases in 2017 or 2018. Another recent study also showed worldwide declines of water storage in endorheic basins in the last 10 more years were caused by limited precipitation with high potential evaporation, which are then intensified by global warming and human activities [19]. Thus, human water use is not the dominant cause but could be a secondary factor of saline lakes' water loss.

Figure 5. GSL monthly water levels in 2016 and the big increases in 2017 and 2018. This figure is plotted on a monthly scale from January 1, 2016 to November 30, 2018, which is calculated and processed using the GSL daily water level records that are available online at http://greatsalt.uslakes.info/Level.asp provided by www.lakesonline.com.

3.6. Correlation between Temperature, Precipitation, Snowfall, River Discharge, And Gsl Water Level

The above analyses have showed apparent temporal patterns of temperature, precipitation, snowfall, and river discharge. Chang and Bonnette have recently examined the correlation between climate and water-related ecosystem services [20]. Here, we use the Pearson correlation coefficient to quantify a general relationship between temperature, precipitation, snowfall, river discharge, and GSL water level, as shown in Table 2. From 1971 to 2016, the significant increasing temperature is significantly and negatively related to GSL water levels. It coincides with the above analysis that local climate warming is a critical variable for water loss. Another significant relationship is between GSL water levels and river discharge. River discharge is significantly and highly correlated with precipitation, as shown in Figure 6, which further indicates that climate factors, especially precipitation along with temperature, are the dominant driving force of GSL water level dynamics.

Table 2. Correlation between temperature, precipitation, snowfall, river discharge, and GSL water level.

	Temperature		Precipitation		Snowfall		River Discharge
	1904–1970	1971–2016	1904–1970	1971–2016	1904–1970	1971–2016	1971–2016
Coefficient	0.19	−0.29	0.20	0.08	0	0.12	0.59
p-value	0.0666	0.0260	0.0521	0.2485	0.4979	0.2089	<0.0001

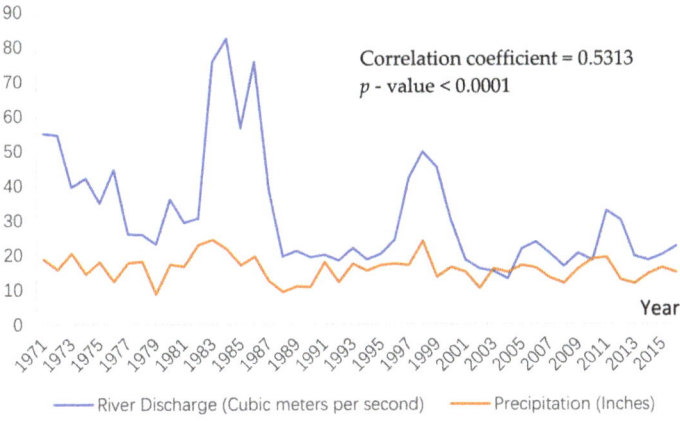

Figure 6. River discharge is highly and significantly related to precipitation.

4. Limitations

Many other studies recently have showed that the GSL water levels are primarily sensitive to climate cycles, given its main outflow is evaporation that is directly changed by lake area and salinity, and precipitation variations mainly drive the GSL water level variation [20–22]. Rather than following the previous research approaches, this proposed study, based on a simple water balance equation of inflow and outflow, explored how climate and weather factors could impact on inflow and outflow, and concluded that climate change and local extreme weather drive the dynamics of GSL water levels. This is not the primary objective of this short communication; we hope this short communication could encourage more high quality studies and inspire scientists to rethink water budget modeling, climate modeling, and hydroclimatological analysis, so that the driving force of water dynamics of salt lakes could be truly modeled and quantified, which then will provide useful and practicable information for water budget planning and water resources management.

We understand that there are many sophisticated models for climate and hydroclimatological analyses, but "all models are wrong, and some are illuminating and useful" [23,24]. Scientists cannot achieve a "correct" model by excessively elaborating modeling procedures and parameters; while for great scientists it is significant to devise simple and evocative models, any overelaboration and over-parameterization is nothing but mediocrity [23,24]. There are some limitations in this short communication. For example, we do not directly examine the impacts of human water use on water level dynamics of GSL, while we analyze the remaining factors for water balance including inflow factors (i.e., precipitation, snowfall, and river discharge) and outflow factors (i.e., evaporation, temperature) that influence the water levels of GSL. We do not have direct measurements of evaporation in the GSL region, but evaporation estimations of water bodies of similar semi-arid regions in Australia are referenced to indicate the impacts of evaporation on GSL water loss. Long-term surveys of evaporation and measurement of human water use can be helpful in order to further analyze and quantify the primary and secondary factors driving the dynamics of saline lakes. Additionally, the 95th and 5th percentiles cannot effectively identify extremely high or low observations of water levels, precipitation, temperature, and snowfall, which further indicates our proposed extremely changing point analysis with mean and two standard deviations is a robust and promising method for hydroclimatological analysis.

5. Conclusion

The proposed temporal trend with the extremely changing point analysis is a promising method to clearly and concisely define and understand the characteristics of extreme climate/weather and their impacts on the declines of saline lakes. Extremes have become foundational information and projections for climate change, which has been highlighted 734 times in the current US Global Change Research Program's Climate Science Special Report [25]. Most often, an explicit definition of extreme is not provided in current research and management, but clearer definitions and quantifications of extremes can support interdisciplinary understanding and decision making of extreme events [6]. Defined by whether an observation is outside its two standard deviations of the mean, the extremely changing points indicate the substantial changes of a variable in its temporal patterns. Using the proposed temporal trend with extremely changing point analysis, this short communication adequately shows that climate change and extreme weather can be the primary driving factors of the dynamics/declines of the Great Salt Lake. Although the impact of one isolated extremely changing point could be limited, two more continuously or clustered extremely changing points can have elevated impacts on the environment. For example, extremely high temperature in 2012, 2014, 2015, and 2016 considerably enhances the continuously increasing evaporation of the GSL since the 1970s. The extremely low precipitation in 1979 is isolated, and therefore its effect was minimized by many of those much higher precipitation observations neighboring and close to it.

Climate changes, especially local warming and extreme weather including both precipitation and temperature, drive the dynamics of the GSL surface levels. Extreme weather, such as extremely high or low precipitation, directly causes the changes in surface levels, and the extremely high temperature in the last five years has resulted in much more water loss through evaporation that can be another main cause of the relatively low surface level in 2016. The increasing temperature trend since the 1970s, as shown in Figure 3, has become a critical role in water loss and hence the decline of the GSL surface levels. As discussed above, many studies have proven that climate warming has resulted in the main water loss through evaporation (i.e., each year about 40% of the total water storage capacity in Australia). The annual increasing rate of 0.0313 in temperature from 1971 to 2016 could result in more than 40% loss of its total water storage each year.

Supplementary Materials: The supplementary material is available online at http://www.mdpi.com/2225-1154/7/2/19/s1.

Funding: This research received no external funding

Conflicts of Interest: The author declares no conflict of interest.

References

1. U.S. Geological Surveys (USGS). Great Salt Lake, Utah. Available online: https://pubs.usgs.gov/wri/wri994189/PDF/WRI99-pdf (accessed on 7 March 2018).
2. Wurtsbaugh, W.A.; Miller, C.; Null, S.E.; DeRose, R.J.; Eilcock, P.; Hahnenberger, M.; Howe, F.; Moore, J. Decline of the world's saline lakes. *Nat. Geosci.* **2017**, *10*, 816. [CrossRef]
3. IPCC. AR5 Climate Change 2014: Impacts, Adaptation, and Vulnerability. Available online: https://www.ipcc.ch/report/ar5/wg2/ (accessed on 14 January 2019).
4. National Centers for Environmental Information; National Oceanic and Atmospheric Administration (NCER-NOAA). Extreme Events. Available online: https://www.ncdc.noaa.gov/climate-information/extreme-events (accessed on 11 May 2018).
5. Hansen, J.; Sato, M.; Ruedy, R. PNAS plus: Perception of climate change. *Proc. Natl. Acad. Sci. USA* **2012**, *109*, E2451. [CrossRef] [PubMed]
6. McPhillips, L.E.; Chang, H.; Chester, M.V.; Depietri, Y.; Friedman, E.; Grimm, N.B.; Kominoski, J.S.; McPhearson, T.; Méndez-Lázaro, P.; Rosi, E.J.; et al. Defining extreme events: A cross-disciplinary review. *Earth's Future* **2018**, *6*, 441–446. [CrossRef]
7. Katz, R.W. Extreme value theory for precipitation: Sensitivity analysis for climate change. *Adv. Water Resour.* **1999**, *23*, 133–139. [CrossRef]
8. Downing, T.E.; Gawaith, M.J.; Olsthoorn, A.A.; Tol, R.S.J.; Vellinga, P. Introduction. In *Climate Change and Risk*; Downing, T.E., Olsthoorn, A.A., Tol, R.S.J., Eds.; Routledge: London, UK, 1999; pp. 1–19.
9. Sokal, R.R.; Rohlf, F.J. Biometry. In *The Principles and Practice of Statistics in Biological Research*, 3rd ed.; Freeman: New York, NY, USA, 1995.
10. Griffiths, D.; Stirling, W.D.; Weldon, K.L. *Understanding Data. Principles and Practice of Statistics*; Wiley: Brisbane, Australia, 1998.
11. Robeson, S.M. Relationship between mean and standard deviation of air temperature: Implications for global warming. *Clim. Res.* **2002**, *22*, 205–213. [CrossRef]
12. Kirono, D.G.C.; Kent, D.M. Assessment of rainfall and potential evaporation from global climate models and its implications for Australian regional drought projection. *Int. J. Climatol.* **2011**, *31*, 1295–1308. [CrossRef]
13. Helfer, F.; Lemckert, C.; Zhang, H. Impacts of climate change on temperature and evaporation from a large reservoir in Australia. *J. Hydrol.* **2012**, *475*, 365–378. [CrossRef]
14. Huntington, T.G. Evidence for intensification of the global water cycle: Review and synthesis. *J. Hydrol.* **2006**, *319*, 83–95. [CrossRef]
15. Johnson, F.; Sharma, A. Measurement of GCM skills in predicting variables relevant for hydroclimatological assessments. *J. Clim.* **2009**, *22*, 4373–4382. [CrossRef]
16. Craig, I.; Green, A.; Scobie, M.; Schmidt, E. *Controlling Evaporation Loss from Water Storages*; Report 1000580/1; National Centre for Engineering in Agriculture: Toowoomba, Austalia, 2005; 207p.
17. Usoskin, I.G.; Solanki, S.K.; Taricco, C.; Bhandari, N.; Kovaltsov, G.A. Long-term solar activity reconstructions: Direct test by cosmogenic 44Ti in meteorites. *Astron. Astrophys.* **2006**, *457*, L25–L28. [CrossRef]
18. Davarzani, H.; Smits, K.; Tolene, R.M.; Illangasekare, T. Study of the effect of wind speed on evaporation from soil through integrated modeling of the atmospheric boundary layer and shallow subsurface. *Water Resour. Res.* **2014**, *50*, 661–680. [CrossRef] [PubMed]
19. Wang, J.; Song, C.; Teager, J.T.; Yao, F.; Famiglietti, J.S.; Sheng, Y.; MacDonald, G.M.; Brun, F.; Schmied, H.M.; Marston, R.A.; et al. Recent global decline in endorheic basin water storges. *Nat. Geosci.* **2018**, *11*, 926–932. [CrossRef] [PubMed]
20. Gillies, R.R.; Chung, O.Y.; Wang, S.Y.S.; DeRose, R.J.; Sun, Y. Added value from 576 years of tree-ring records in the prediction of the Great Salt Lake level. *J. Hydrol.* **2015**, *529*, 962–968.
21. Mohammed, I.N.; Tarboton, D.G. An examination of the sensitivity of the Great Salt Lake to changes in inputs. *Water Resour. Res.* **2012**, *48*. [CrossRef]
22. Wang, S.Y.; Gillies, R.R.; Jin, J.; Hipps, L.E. Coherence between the Great Salt Lake level and the Pacific quasi-decadal oscillation. *J. Clim.* **2010**, *23*, 2161–2177.
23. Box, G.E.P. Science and Statistics. *J. Am. Stat. Assoc.* **1976**, *71*, 791–799. [CrossRef]

24. Box, G.E.P. *Robustness in the Strategy of Scientific Model Building*; Launer, R.L., Wilkinson, G.N., Eds.; Robustness in Statistics; Academic Press: New York, NY, USA, 1979; pp. 201–236.
25. USGCRP (The U.S. Global Change Research Program). *Climate Science Special Report: Fourth National Climate Assessment*; Wuebbles, D.J., Fahey, D.W., Hibbard, K.A., Dokken, D.J., Stewart, B.C., Maycock, T.K., Eds.; U.S. Global Change Research Program: Washington, DC, USA, 2017; Volume I, 470p. [CrossRef]

© 2019 by the author. Licensee MDPI, Basel, Switzerland. This article is an open access article distributed under the terms and conditions of the Creative Commons Attribution (CC BY) license (http://creativecommons.org/licenses/by/4.0/).

Article

Temporal Changes in Precipitation and Temperature and their Implications on the Streamflow of Rosi River, Central Nepal

Ngamindra Dahal [1], Uttam Babu Shrestha [2,*], Anita Tuitui [3] and Hemant Raj Ojha [4]

1. South Asia Institute of Advanced Studies (SIAS), Min Bhawan-34, N.K.Singh Marga 306, Kathmandu 44600, Nepal; ngamindra@gmail.com
2. Institute for Life Sciences and the Environment, University of Southern Queensland, Toowoomba 4350, Australia
3. Central Department of Hydrology and Meteorology, Tribhuvan University, Kathmandu 44618, Nepal; go4uranu@gmail.com
4. Institute for Studies and Development Worldwide (IFSD) and University of Canberra, Sydney 2000, Australia; ojhahemant1@googlemail.com
* Correspondence: ubshrestha@yahoo.com; Tel.: +61-481-098-529

Received: 23 October 2018; Accepted: 21 December 2018; Published: 28 December 2018

Abstract: Nepal has experienced recent changes in two crucial climatic variables: temperature and precipitation. Therefore, climate-induced water security concerns have now become more pronounced in Nepal as changes in temperature and precipitation have already altered some hydrological processes such as the river runoff in some river systems. However, the linkage between precipitation patterns and streamflow characteristics are poorly understood, especially in small rivers. We analysed the temporal trends of temperature, precipitation, and extreme indices of wet and dry spells in the Rosi watershed in Central Nepal, and observed the temporal patterns of the streamflow of the Rosi river. We also examined the linkages between the average and extreme climate indices and streamflow. We found that the area has warmed up by an average of 0.03 °C/year, and has seen a significant decline in precipitation. The dry spell as represented by the maximum length of the dry spell (CDD) and the magnitude of dryness (AII) has become more pronounced, while the wet spell as represented by the number of heavy rainfall days (R5D) and the precipitation intensity on wet days (SDII) has diminished significantly. Our analysis shows that recent changes in precipitation patterns have affected the streamflow of the Rosi river, as manifested in the observed decline in annual and seasonal streamflows. The decrease in the availability of water in the river is likely to have severe consequences for water security in the area.

Keywords: Himalaya; streamflow; extreme rainfall; watershed

1. Introduction

The impact of climate change on water availability is a major concern worldwide, and the question of how water systems remain resilient under changing climate conditions has dominated the world's science and policy agenda recently [1,2]. Such a climate-induced water security concern is nowhere more visible than in the Himalayan region. Climate change has significantly impacted the glaciers and water resources in the Himalayan region, which is the water tower of Asia that provides water and related hydrological services to 1.3 billion people downstream, from Afghanistan in the west to Vietnam in the east [3,4]. The melting of snow and glaciers is a significant hydrological process in this region that sustains the flows of rivers during the dry season [3], and this crucial hydrological process is being affected by climate change, particularly regarding changes in temperature and precipitation [5]. Furthermore, rising temperature and changes in precipitation alter some components of hydrological systems such as precipitation extremes, increasing evaporation, and changes in river runoff [6,7]. Due to these climate-induced changes, two major impacts on hydrological systems are expected to escalate. First, the availability of water and related hydrological services are likely to decrease due to the recession of glaciers in the Himalaya [8]. Second, climate hazards such as flood and drought due to precipitation extremes are expected to increase with climate change [7]. Such fundamental alterations in the hydrological regime, which are attributed primarily to climate change, will have a cascading impact on the irrigated agriculture and installed hydropower capacity, as well as the biodiversity and natural resources [9], and will eventually intensify the regional conflicts in this region [10].

When aggregated at the national level, water is one of the most abundant natural resources in the Himalayan country of Nepal [11]. Most of the rivers of Nepal are snow and glacier-fed, and the melting of snow and glaciers provides sustained flows during dry seasons [12,13]. However, the narrative of a national-scale water surplus hides the stark reality of many localities and regions facing acute water shortages [14]. A higher rate of warming than the global average [11,15], erratic rainfall with a greater spatial and temporal variability [16], and a prolonged drought spell [17] have been reported recently in Nepal, which clearly indicate the growing impact of climate change in the country. Consistent with these research findings, Nepal is already considered the 14 most vulnerable country in the world in terms of the climate change vulnerability index [18]. As reported elsewhere, the observed and predicted changes in the climate are likely to alter Nepal's hydrological systems. Combined with the rapid land-use transformations taking place across many of the mountain landscapes of Nepal, climate change is poised to escalate water insecurity in many water-deficit regions of the country, including several hilltop cities such as Dhulikhel in the central Nepal Himalaya, which could have severe consequences for the amount and seasonality of water availability. The effect of water scarcity has already been pronounced in many villages due to the drying up of local water sources that have, in some instances, created competitions and conflicts [19], as well as forced migration [20].

Streamflow is an important hydrological variable that can be used as an indicator of hydrological responses to climate change and variability [21–23]. It is determined by catchment heterogeneity (land use, anthropogenic water usage) along with hydroclimatic processes such as precipitation, temperature, infiltration, and evapotranspiration [4,24]. Climate change or variability contribute to the increased variability of stream runoff due to changes in the timing, frequency, and intensity of precipitation events [13]. Therefore, analysing streamflow trends in watersheds and identifying the causes and drivers of changes has been a focus of hydrological research globally [21,23], including in Nepal [24,25]. In recent decades, runoff changes in Nepalese rivers have been reported as being associated with the effects of climate change [11,24]. Fluctuations in the natural streamflow affect water availability, which has direct consequences on the livelihood of the people who are heavily dependent on streamflow for agriculture. It also has a potential impact on the economic development of the country, whose economy largely depends on agriculture and hydropower development.

Although climate change is a global phenomenon, it has noticeable local impacts affecting local biodiversity, ecosystems, water availability, and livelihoods [9,15]. Several efforts have been made recently to analyse national and regional patterns of climate change in Nepal [11,16,17] and their impacts on hydrology [26–29]. However, very few studies have focused on local watersheds, despite the importance of small and localized watersheds to local livelihoods, including ecosystem goods and services to the predominantly agrarian society in Nepal. Most of the hydrological studies have been conducted in the larger river basins of Nepal such as the Dudhkoshi [12], Indrawati [28], Koshi [29–31], Bagmati [32–34], Gandaki [26], Tamor and Seti [35], and Karnali [27] river watersheds. These studies, which were conducted at the national level or in the large river basins, analysed the streamflow trends of various rivers in Nepal [24], but little work has been done towards assessing the link between changing climate and streamflow characteristics. This paper presents the findings of a study conducted in a small watershed (of Rosi stream, about 30 km east of the capital city of Kathmandu) with the goal of understanding the changes in hydroclimatic dynamics, including the analysis of the potential association between streamflow and climatic parameters. The streamflow dynamic in Rosi is directly linked to the availability of water for anthropogenic usage such as drinking.

The Rosi watershed supplies drinking water to three municipalities in the Kavre district in central Nepal: namely, Panauti, Banepa, and Dhulikhel. For the last few decades, a large part of Rosi water has been diverted to cater to the drinking water needs of Dhulikhel, and more recently to the two other cities, too. The Rosi-based water supply scheme also has a history of conflict and cooperation between upstream rural and downstream urban communities, which are themselves undergoing rapid change in relation to urbanisation, livelihood trajectories, and farming practices. The results of this study could be useful to the similar watersheds of Nepal and in other mountainous countries. The three specific objectives of this study are to: (a) investigate the spatial and temporal trends of temperature and precipitation as well as moderate extreme indices related to precipitation; (b) analyse the temporal streamflow trends of the Rosi river based on the available dataset (from 1971 to 2014); and (c) assess the linkage between the river discharge with precipitation parameters in the Rosi river watershed.

2. Materials and Methods

2.1. Study Area

This study focuses on the Rosi watershed, which is located in the western part of the Kavrepalanchok District of Nepal (Figure 1) and covers approximately 87 km^2 area out of 540 km^2 of its entire basin [36]. The Rosi watershed is one of the sub-basins of Sunkoshi river, which is a tributary of the Kosi river that flows from the north to the south of India. The watershed extends from the latitudes between 27°22' and 27°42' N, and longitudes between 85°22' and 85°48' E, with an altitudinal range from 1450 m to 2828 m. The area has a sub-tropical climate with annual temperature ranges between 9–24° Celcius, and receives moderate annual rainfall of 1040–2225 mm. The Rosi watershed can be characterised as a typical watershed in the mid-hills of Nepal, which are dominated by community forests, fragmented small-scale agriculture lands, scattered settlements, and small towns. This watershed provides various ecosystem goods and services to local people, particularly hydrological services, including drinking water to the approximately 50,000 inhabitants of the Panauti, Banepa, and Dhulikhel municipalities. To harness the water from Rosi, a large-scale project called the Kavre Valley integrated water supply project with financial and technical cooperation from the Asian Development Bank has been implemented since 2013.

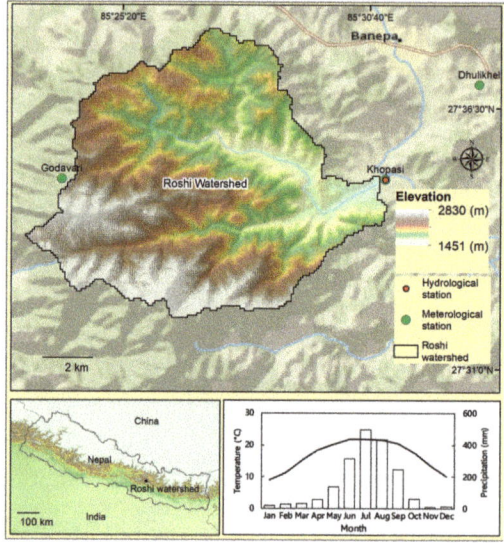

Figure 1. Map of the study area showing average monthly temperature and precipitation (climatology) calculated for the period between 1971–2014.

2.2. Data

The historical meteorological and hydrological data for the Rosi river basin were collected from the Department of Hydrology and Meteorology (DHM), Nepal. There are three meteorological stations (Kophasi, Dhulikhel: index number 1024, and Godavari: index number 1022) located in and around the watershed boundary area (Figure 1). However, temperature data were available only for the Godavari meteorological station. The meteorological data from the year 1971 to 2014 that were used in this analysis include the daily precipitation, as well as the minimum and maximum temperature. The gauge flow data measured at Panauti gauging station (index number 1049) was also collected from the DHM. The hydrological station in Panauti and meteorological station in Kophasi were located within a kilometer distance.

2.3. Data Analysis

We calculated trends of annual, seasonal, and maximum (extreme) discharge on the streamflow. We conducted a regression analysis to identify the trends in the river discharge. We also compared the spatial variation in precipitation trends among three stations, and analysed temporal changes in the precipitation and temperature trends using linear regression. Seasonal analysis was based on four seasons: winter (December–February), pre-monsoon (March–May), monsoon (June–September), and post-monsoon (October–November). To understand the climate dynamics, moderate climate extreme indices that describe events with short return periods are appropriate [37]. Currently, 27 different climate extreme indices were suggested by CCl/CLIVAR (Commission for Climatology/Climate and Ocean: Variability, Predictability and Change)/JCOMM (Joint World Meteorological Organization(WMO)-Intergovernmental Oceanographic Commission(IOC), Technical Commission for Oceanography and Marine Meteorology Expert Team (ET) on Climate Change Detection and Indices (ETCCDI) [38]. We selected six indices; three were related to the dry spell (CDD, maximum length of the dry spell; FDD, the number of dry spells; and AII, the magnitude of dryness) and three were related to the wet spell (SDII, the precipitation intensity on wet days; R5D, the number of heavy rainfall days; and R20, the frequency of extremely heavy precipitation) for moderate precipitation extremes. These indices represent both the intensity and duration of dry and wet spells and are directly related to the streamflow. The details of those indices are given in Table 1. We used the Mann–Kendall test [39,40] for detecting trends in temperature, precipitation, precipitation extremes, and streamflow. The Mann–Kendall test is a non-parametric test that is used to identify a trends in time-series data such as precipitation and temperature. This test is a widely-used test to detect significant trends in hydroclimatic data [16,21,27,29]. This test is not affected by the actual distribution of the data and is less sensitive to outliers. Therefore, it is more suitable for detecting trends in climatic and hydrological data, which are usually skewed, and may contain outlier observations [41]. We used Sen's non-parametric estimate of the slope to determine the magnitude of trends [42], as the Mann–Kendall test can examine the time series trend, but not the extent. The relationship between precipitation parameters (annual and monthly averages, and precipitation extremes) and discharge was calculated using Pearson's correlation.

Table 1. Description of the indices.

Indices	Name	Definition	Method of Calculation	Unit
R5D	Number of heavy rainfall days	Annual count of days when days rainfall ≥ five mm	$RR_{ij} \geq$ five mm	Days
R20	Number of very heavy rainfall days	Annual count of days when days rainfall ≥ 20 mm	$RR_{ij} \geq$ 20 mm	Days
SDII	Simple daily intensity index	Annual mean rainfall when Precipitation ≥ one mm		Days
CDD	Maximum length of dry spell	Maximum number of consecutive days with RR < one mm	$RR_{ij} <$ one mm	Days
FDD	Number of dry spells	Consecutive period with at least eight dry days	R < one mm	Frequency
AII	Aridity index	Ratio between the total rain on dry days and the number of dry days	Total rain on days with (R < 10 mm)/number of days with R < 10 mm	mm

RR is the daily precipitation amount on the day in a period. R5D is the number of heavy rainfall days. R20 is the frequency of extremely heavy precipitation. SDII is the precipitation intensity on wet days. CDD is the maximum length of the dry spell. FDD is the number of dry spells. AII is the magnitude of dryness.

3. Results

3.1. Spatial and Temporal Patterns of Precipitation

Monthly precipitation patterns (averaged over the study period from 1971–2014) of the three meteorological stations are given in Figure 2. The Rosi river watershed received the majority of precipitation (~80%) mainly during the monsoon season with little spatial variation (76% in Khopasi, 81% in Godavari, and 79% in Dhulikhel). Khopasi was the driest, while Godavari was the wettest station. July was the wettest month, whereas November was the driest month in the study area. Overall, the annual rainfall in the two stations (Godavari and Dhulikhel) of the study area significantly decreased, with a rate of −10.4 mm/year ($p = 0.006$) in Godavari and −9.1 mm/year ($p = 0.010$) in Dhulihel (Figure 3). The seasonal rainfall pattern showed no significant trends except in monsoon season (Table 2). There was a significant decrease in monsoon rainfall across all three stations and a maximum decrease occurred in Godavari, with −10.3mm/year ($p = 0.002$) and a minimum in Khopasi with −6.3mm/year ($p = 0.046$).

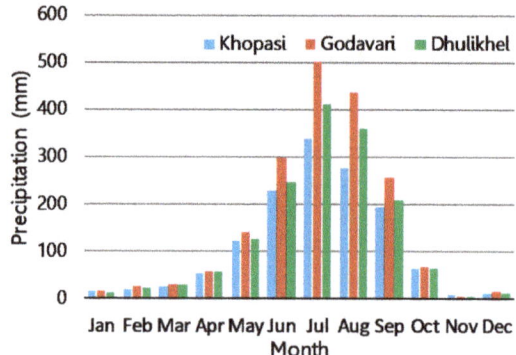

Figure 2. Variations in averaged (1971–2014) monthly precipitation patterns in three meteorological stations.

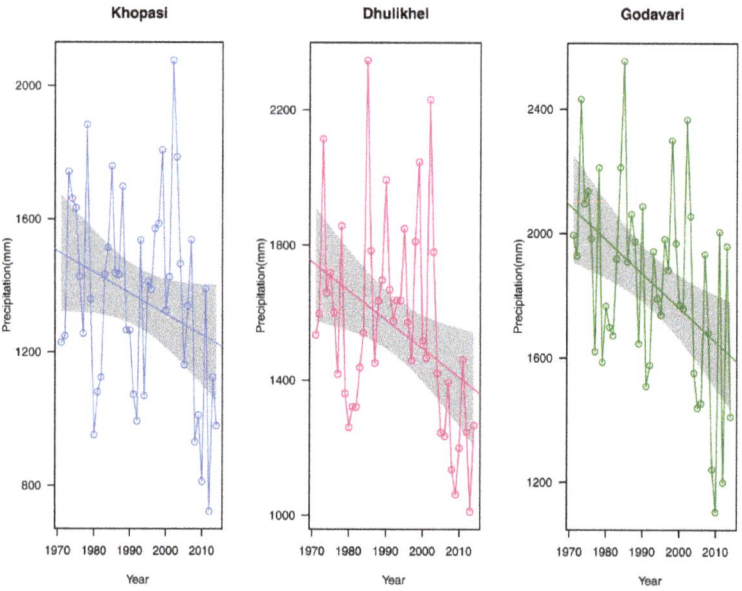

Figure 3. Temporal trends of annual precipitation in three meteorological stations. The grey shadows represent the standard error of the regression line.

Table 2. Trends in the annual and seasonal precipitation.

Weather Stations	Annual	Pre-Monsoon	Monsoon	Post-Monsoon	Winter
Khopasi	−6.296	0.300	−6.260 *	−0.386	−0.215
Godavari	−10.435 *	0.501	−10.358 *	−0.207	−0.117
Dhulikhel	−9.122 *	−0.349	−7.120 *	−0.922	−0.162

* $p = < 0.05$.

3.2. Trends in Precipitation Extremes

Overall, the dry spell in the study area was increasing, while the wet spell was decreasing (Figure 4). Three wetness indices (SDII, R5D, and R20) showed a significant decrease in the study area, suggesting that the duration and intensity of the heavy precipitation events in the area have declined over the study period. SDII, which measures the precipitation intensity on wet days, showed a significant decreasing trend in Khopasi (−0.047/year, $p = 0.040$). The Godavari had a maximum decrease (−1.451/year, $p = 0.018$) in the number of heavy rainfall days (R5D), and Dhulikhel had a maximum decrease (0.167/year, $p = 0.038$) in the frequency of extremely heavy precipitation, R20 (Table 3).

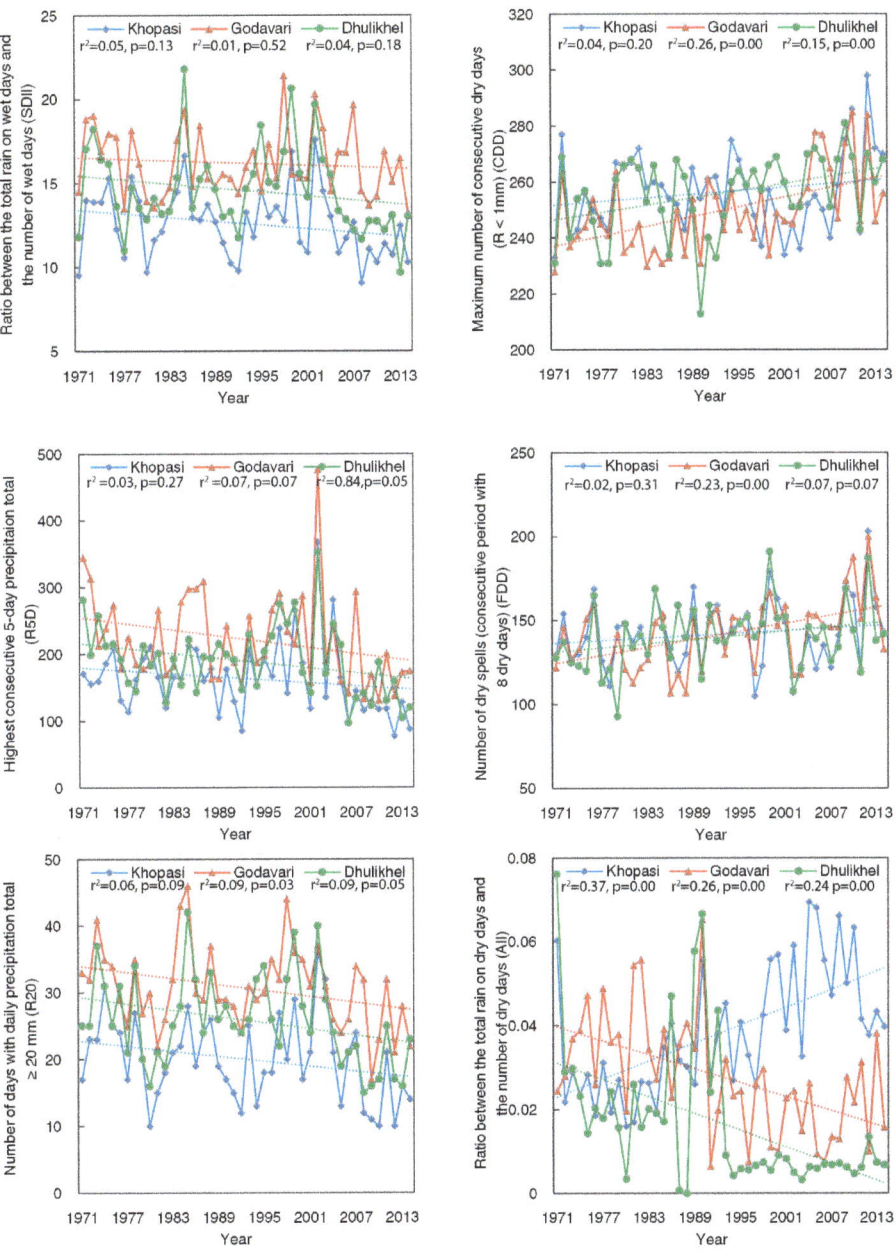

Figure 4. Trends of different wet and dry indices.

AII, CDD, and FDD are dryness indices that are used to study trends in the dry spell. The magnitude of dryness (AII) has significantly increased, albeit at a small rate in Khopasi and Dhulikhel, but decreased in Godavari. A maximum number of consecutive dry days as measured by CDD (maximum length of the dry spell) significantly increased in Godavari and Dhulikhel. A significant increase in FDD (number of dry spells) was observed only in Godavari.

Table 3. Trends (Sen's slope) in the dryness and wetness indices.

	Khopasi	Godavari	Dhulikhel
AII	0.0007 ***	−0.0005 ***	0.0004 ***
CDD	0.0001	0.148 ***	0.537 ***
FDD	0.250	0.707 **	0.375
SDII	−0.047 **	−0.018	−0.041 *
R5D	−1.125 **	−1.452 **	−1.363 **
R20	−0.143 *	−0.149 *	−0.167 **

*** = 0.001, ** = < 0.05, * = < 0.10

3.3. Temporal Patterns of Temperature

The annual mean, maximum, and minimum temperature recorded only at the Godavari station were analysed (Figure 5), as the data were available only to this station. The annual mean temperature trend showed that the area warmed up by 0.03 °C/year ($p = <0.0001$). The maximum temperature in the study area increased at a rate of 0.067 °C/year, and the minimum temperature increased at a rate of 0.005 °C/year over the last 44 years. The increasing trend in mean annual temperature over the past four decades is consistent with the national and global averages.

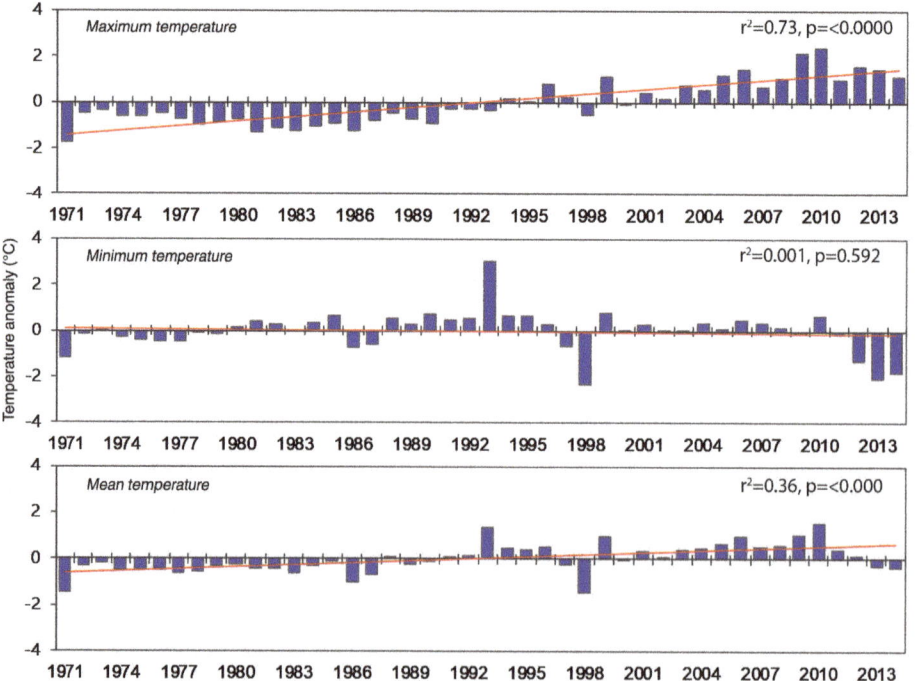

Figure 5. Mean, maximum, and minimum temperature anomalies with respect to 1971–2014.

3.4. Hydrological Change and Its Linkage with Precipitation Indices

The annual river discharge pattern from 1971 to 2014 indicates a gradual decline of the river flow, and the trend is statistically significant at the 10% significance level based on the Mann–Kendall test (Figure 6). The trend slope based on Sen's method showed that the rate of decrease in the mean annual discharge of the Rosi river was $-0.015 \text{m}^3/\text{s/year}$ ($p = 0.08$) over the last 44 years. The monthly average discharge reached a maximum in August and a minimum in December (Figure 7). Seasonally, the maximum flow of river occurred during the monsoon season (June–September), whereas the minimum flow occurred during the winter (December–February). In the post-monsoon season (October–November), the flow in the river is sustained by the infiltration supply available in the monsoon season. The trends of seasonal flows in the winter, monsoon, and post-monsoon showed negative trends except for the pre-monsoon, and a positive but statistically significant trend was found only in the streamflow in the monsoon season ($-0.041 \text{m}^3/\text{s/year}$, $p = 0.04$).

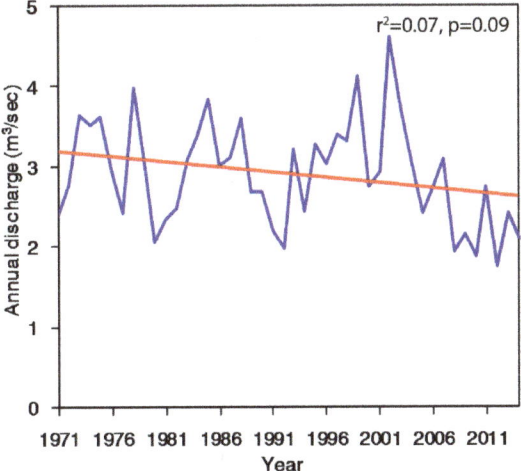

Figure 6. Trends of annual discharge.

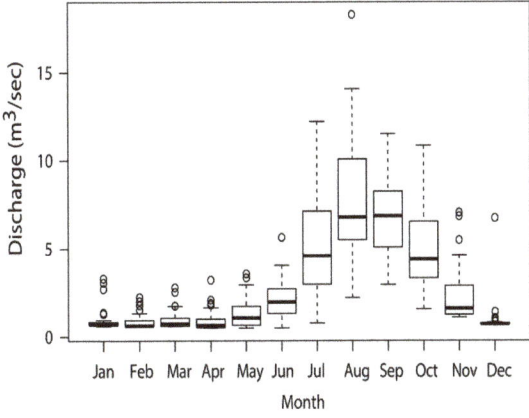

Figure 7. Monthly average of the discharge calculated for the period between 1971–2014.

The annual streamflow in the Rosi river was highly correlated with annual precipitation. Figure 8 illustrates a high correlation between the annual precipitation and the runoff in the Rosi river (Pearson correlation (r) = 0.83, $p = < 0.001$). Not only the annual precipitation and annual runoff were significantly correlated, a strong correlation between the runoff and the dry spell and wet spell indices was also observed in the study area. CCD and FDD were negatively correlated with discharge, while SDII, R5D, and R20 showed a significant positive correlation, suggesting that the flow of the river was highly dependent on the extreme precipitation events (Figure 9). We did not find any significant correlation between the annual temperature and the annual discharge, indicating that the increasing temperature in the Rosi watershed may have had a minimum role in the streamflow.

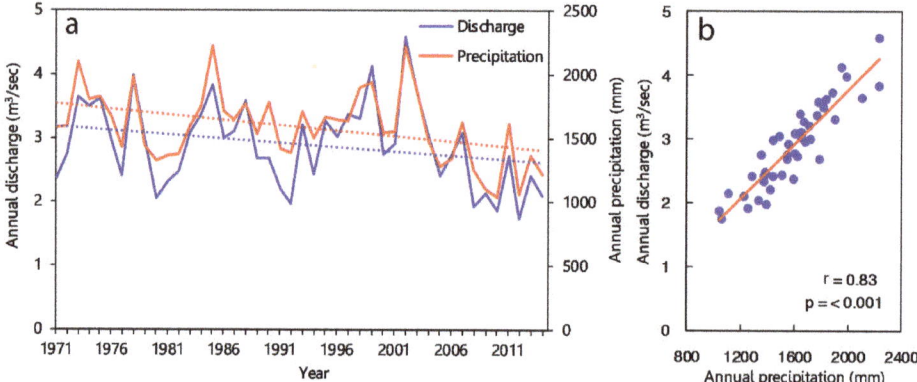

Figure 8. (a) Temporal trends of annual discharge and annual precipitation; (b) correlation between annual discharge and annual precipitation.

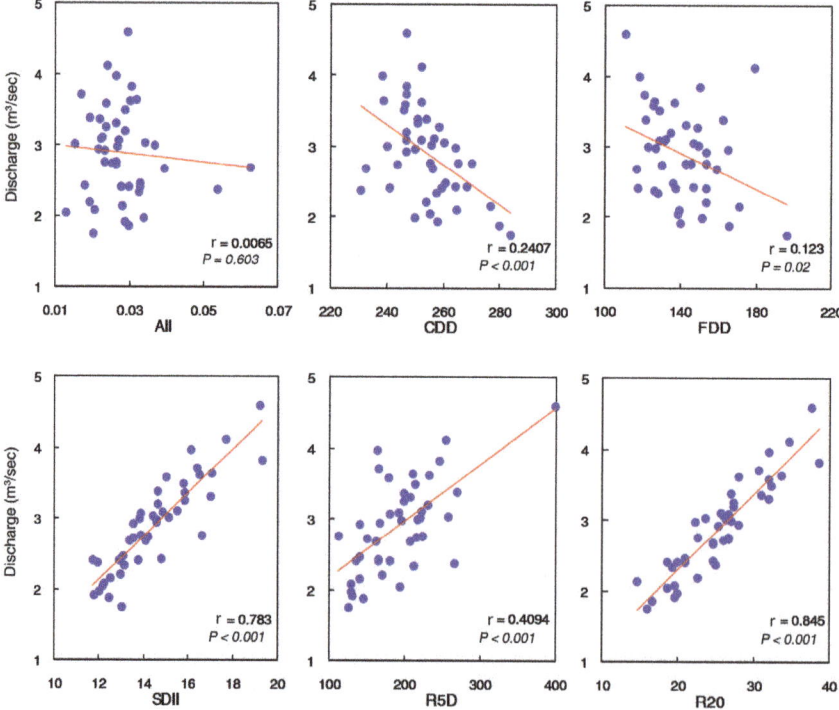

Figure 9. Correlation of annual discharge with different wet and dry spells.

4. Discussion

We analysed runoff trends and assessed various climatic drivers of the runoff change, along with the trends in the temperature, precipitation, and extreme precipitation indices. Our results of the decreasing annual precipitation and increasing warming and precipitation extremes resonate with the results that have been reported in previous studies [16,17,27]. Although, at the national level, the precipitation pattern has remained more or less stable, a large spatial variability in precipitation has been observed across various localities in Nepal [16]. In western Nepal's Karnali basin, for example, the annual precipitation is following a declining trend [27], while in the Gandaki Basin in central Nepal, annual rainfall has remained stable [26]. The seasonality and magnitude of rainfall were found to be more or less constant in the high mountain areas of Nepal [43]. We did not observe any trends in seasonal rainfalls except in the monsoon, which was declining, although some studies [26] reported a significant increase in monsoon rainfall, while post-monsoon, pre-monsoon, and winter rainfalls were decreasing. This indicates not only spatial heterogeneity in the precipitation pattern, but also a seasonal variation in the precipitation trends in different localities of Nepal.

Our results of an increasing dry spell (the number of consecutive dry days) and decreasing wet spell (the number of rainy days) are consistent with the findings of Karki et al. (2017), who reported a significant positive trend in the number of consecutive dry days and a significant negative trend in the number of rainy days. This implies the prolongation of the dry spell of the study area. More importantly, our finding reinforces the commonly reported experience of the local people of Nepal. In a recent national survey, about 86.1% (n = 5060) of the respondents reported that they had experienced drought, and 99.3% respondents reported increasing drought over the past 25 years [44]. Likewise, our findings on warming trends are similar to those reported in previous studies that have been conducted in other areas of the country [11,43,45].

On a national-level analysis, both increasing (59%) and decreasing (41%) streamflow trends were found in Nepal [24]. Our result of a decline in streamflow parallels with the streamflow in the Bagmati river [32], but contradicts with the streamflow of the Jhikhu river, where an upward trend of streamflow was reported [24]. The streamflows of Nepalese rivers are determined by several factors such as the quantum of melting snow, glacier, groundwater, and precipitation [12,24,27]. In the small and non-snowfed rivers such as Bagmati and Jhikhu, streamflow is highly dependent on precipitation events [24,32], which means that runoff events in the mid-hills catchments of such small rivers are closely correlated to the rainfall intensity [46].

In the glacier-fed or snowfed river, temperature and discharge are correlated, as an increasing temperature accelerates the ablation process [47]. The role of temperature in non-snowfed river system such as that of the Rosi river is expected to be minimal; therefore, as shown in our results, temperature has a minimum role in the river discharge. Increasing temperature causes the melting of snow and glaciers, enhancing river flows noticeably in the pre-monsoon and winter season in snowfed rivers [24]. In such snowfed river systems, the role of temperature is more pronounced than the role of precipitation. For example, the streamflow in the snowfed Karnali river remains constant with a decreasing precipitation trend [27]. A significant increase in the annual runoff with increasing air temperature and decreasing precipitation was observed on the Tibetan side of the Himalayan region [48]. Therefore, global phenomena such as increasing temperature might have a lesser effect in a non-snowfed river such as Rosi compared to snowfed river systems.

Along with precipitation, water availability in smaller watersheds is prone to be impacted by changes in land use and land management [49]. In some instances, anthropogenic activities such as irrigated agriculture and population increase have significantly shaped the runoff in the streams [50] as water is drawn from the river for anthropogenic usage. In the middle mountains of Nepal, reforestation in the hillsides is reported to cause a considerable amount of water loss through evapotranspiration, contributing to the observed decline in seasonal streamflow [51]. Globally, the impact of reforestation or afforestation on local water yield was found to be negative; additional forest cover reduces downstream water availability [52]. In the Rosi watershed, 8.19 km^2 area of forest was added (37% increase) from

1976 to 2014, with a decrease in the agricultural and barren land [53]. In the headstream of Rosi, there are several small streams (Muldole Khola, Khar Khola, Gudgude Khola, Bairamahadev Kholsi, and Shishakhani Kholsi), and most of the headstream is under forest cover. Due to the difficult and steep terrain, water usage in the upstream areas is very low. However, our research does not make a final claim between forest condition and streamflow, as there is a range of other factors, such as stone quarrying, agricultural intensification, and others are at play in the catchment, all collectively impacting the runoff. Therefore, changes in land use, particularly a recent increase in the forest in the upper part of the watershed, might have contributed to the decline in runoff downstream by increasing the use of water for evapotranspiration and reducing infiltration.

This study shows that the Rosi watershed has undergone changes in climatic parameters. We also observed that the region has undergone significant land-use transformation. The area has warmed up and received less rainfall, with increasing dry spells and decreasing wet spells. Both the change in climatic factors—particularly precipitation—and local land use have impacted the streamflow of the Rosi river, leading to a decline in the annual and seasonal discharges. While our analysis demonstrates a significant causal association between precipitation parameters and streamflow trends, this study has not been able to segregate the impact of land-use change in the upstream regions. Our fieldwork and interviews with local seniors conducted between November 2014 and May 2016 confirm their experience of receiving increasingly less rainfall in recent years, and that the condition of forest in the watershed has improved. Therefore, a more detailed study is needed to understand the role of land use and management, particularly regarding how the increased area of forest affects water availability in the downstream of the Rosi river.

Considering the projected impact of climate change, proper adaptation strategies and plans need to be formulated and implemented in order to cope with the escalating water insecurity challenge in the study area. The strategy for securing water access should also incorporate measures to monitor the impact of changing land use, including how increasing forest areas affect the hydrological cycle in the small watershed of the Rosi river. Such monitoring of the forest–hydrology relationship is particularly important, given the contradictory scientific claims being made on the link [52]: it is widely anticipated that forests play a crucial role in water recharge; however, the opposite was true in many cases [51]. While this study has generated some insights into the relationship among the changes in precipitation and hydrological systems, a robust monitoring system needs to be established in order to help the adaptive management of water and watersheds to secure the water future of the region. Neither national nor a large river basin-based approach can generate the evidence that is needed to ensure locality-specific sustainable water management systems in the Himalayan region.

5. Conclusions

The decline of streamflow would cause a shortage of water for domestic, agricultural, and industrial uses in the downstream. If the current trend of declining water flows and increasing population growth continue, water insecurity will be exacerbated. The decline in the availability of water, especially in the dry season will have severe consequences for water security in the Rosi Valley. While the national level studies showed a vast heterogeneity in precipitation and streamflow patterns, local research such as this provides vital evidence to inform local-level water management planning.

Author Contributions: N.D.—Study design, data collection and manuscript writing, U.B.S.—Data analysis, manuscript writing, A.T.—Data analysis, manuscript writing, H.R.O.—Study design, manuscript writing.

Funding: UK Ecosystem Services and Poverty Alleviation (ESPA) supported project "The political economy of water Security, ecosystem services and livelihoods in the Western Himalayas" and Canadian International Development Research Center funded project "Climate Adaptive Water Management Plans for Cities in South Asia".

Acknowledgments: The authors gratefully acknowledge the contribution made by Kamal Devkota and Kaustuv Raj Neupane during data collection. The authors also acknowledge UK Ecosystem Services and Poverty Alleviation (ESPA) supported project "The political economy of water Security, ecosystem services and livelihoods in the Western Himalayas". We have benefitted from collaboration with Bhaskar Vira and Eszter Kovach,

Cambridge University. The analysis of data and follow up study for this paper was supported by Canadian International Development Research Center funded project "Climate Adaptive Water Management Plans for Cities in South Asia".

Conflicts of Interest: No conflict of Interest.

References

1. Vörösmarty, C.J.; Green, P.; Salisbury, J.; Lammers, R.B. Global water resources: Vulnerability from climate change and population growth. *Science* **2000**, *289*, 284–288. [CrossRef] [PubMed]
2. Arnell, N.W.; van Vuuren, D.P.; Isaac, M. The implications of climate policy for the impacts of climate change on global water resources. *Glob. Environ. Chang.* **2011**, *21*, 592–603. [CrossRef]
3. Immerzeel, W.W.; Droogers, P.; De Jong, S.M.; Bierkens, M.F.P. Large-scale monitoring of snow cover and runoff simulation in Himalayan river basins using remote sensing. *Remote Sens. Environ.* **2009**, *113*, 40–49. [CrossRef]
4. Li, H.; Xu, C.Y.; Beldring, S.; Tallaksen, L.M.; Jain, S.K. Water resources under climate change in Himalayan basins. *Water Resour. Manag.* **2016**, *30*, 843–859. [CrossRef]
5. Barnett, T.P.; Adam, J.C.; Lettenmaier, D.P. Potential impacts of a warming climate on water availability in snow-dominated regions. *Nature* **2005**, *438*, 303. [CrossRef] [PubMed]
6. Xu, H.; Taylor, R.G.; Xu, Y. Quantifying uncertainty in the impacts of climate change on river discharge in sub-catchments of the Yangtze and Yellow River Basins, China. *Hydrol. Earth Syst. Sci.* **2011**, *15*, 333–344. [CrossRef]
7. Bolch, T.; Kulkarni, A.; Kääb, A.; Huggel, C.; Paul, F.; Cogley, J.G.; Frey, H.; Kargel, J.S.; Fujita, K.; Scheel, M.; et al. The state and fate of Himalayan glaciers. *Science* **2012**, *336*, 310–314. [CrossRef]
8. Immerzeel, W.W.; Van Beek, L.P.; Bierkens, M.F. Climate change will affect the Asian water towers. *Science* **2010**, *328*, 1382–1385. [CrossRef]
9. Xu, J.; Grumbine, R.E.; Shrestha, A.; Eriksson, M.; Yang, X.; Wang, Y.U.N.; Wilkes, A. The melting Himalayas: Cascading effects of climate change on water, biodiversity, and livelihoods. *Conserv. Biol.* **2009**, *23*, 520–530. [CrossRef]
10. Akhtar, M.; Ahmad, N.; Booij, M.J. The impact of climate change on the water resources of Hindukush–Karakorum–Himalaya region under different glacier coverage scenarios. *J. Hydrol.* **2008**, *355*, 148–163. [CrossRef]
11. Shrestha, A.B.; Aryal, R. Climate change in Nepal and its impact on Himalayan glaciers. *Reg. Environ. Chang.* **2011**, *11*, 65–77. [CrossRef]
12. Nepal, S.; Krause, P.; Flügel, W.A.; Fink, M.; Fischer, C. Understanding the hydrological system dynamics of a glaciated alpine catchment in the Himalayan region using the J2000 hydrological model. *Hydrol. Process.* **2014**, *28*, 1329–1344. [CrossRef]
13. Pradhananga, N.S.; Kayastha, R.B.; Bhattarai, B.C.; Adhikari, T.R.; Pradhan, S.C.; Devkota, L.P.; Shrestha, A.B.; Mool, P.K. Estimation of discharge from Langtang River basin, Rasuwa, Nepal, using a glacio-hydrological model. *Ann. Glaciol.* **2014**, *55*, 223–230. [CrossRef]
14. Ojha, H. High and Dry. Kathmandu Post, 21 October 2016. Available online: http://kathmandupost.ekantipur.com/news/2016-10-21/high-and-dry-20161021081608.html (accessed on 14 June 2017).
15. Shrestha, U.B.; Gautam, S.; Bawa, K.S. Widespread climate change in the Himalayas and associated changes in local ecosystems. *PLoS ONE* **2012**, *7*, e36741. [CrossRef] [PubMed]
16. Duncan, J.M.; Biggs, E.M.; Dash, J.; Atkinson, P.M. Spatio-temporal trends in precipitation and their implications for water resources management in climate-sensitive Nepal. *Appl. Geogr.* **2013**, *43*, 138–146. [CrossRef]
17. Karki, R.; Schickhoff, U.; Scholten, T.; Böhner, J. Rising precipitation extremes across Nepal. *Climate* **2017**, *5*, 4. [CrossRef]
18. Eckstein, D.; Kunzel, V.; Schafer, L. *Global Climate Risk Index 2018*; Germanwatch e.V.: Bonn, Germany, 2017; Available online: https://germanwatch.org/sites/germanwatch.org/files/publication/20432.pdf (accessed on 27 December 2018).
19. The Himalayan Times. Save Water Sources (Editorial). 20 March 2017. Available online: https://thehimalayantimes.com/opinion/editorial-save-water-sources/ (accessed on 16 June 2017).

20. Bernet, D. Fleeing Drought D+C Development and Cooperation. 29 April 2013. Available online: https://www.dandc.eu/en/article/climate-change-nepal-entire-villages-must-relocate-because-water-scarcity-getting-worse (accessed on 16 June 2017).
21. Tao, H.; Gemmer, M.; Bai, Y.; Su, B.; Mao, W. Trends of streamflow in the Tarim River Basin during the past 50 years: Human impact or climate change? *J. Hydrol.* **2011**, *400*, 1–9. [CrossRef]
22. Fu, G.; Charles, S.P.; Viney, N.R.; Chen, S.; Wu, J.Q. Impacts of climate variability on stream-flow in the Yellow River. *Hydrol. Process. Int. J.* **2007**, *21*, 3431–3439. [CrossRef]
23. Zhao, J.; Qiang, H.; Jianxia, C.; Dengfeng, L.; Shengzhi, H.; Xiaoyu, S. Analysis of temporal and spatial trends of hydro-climatic variables in the Wei River Basin. *Environ. Res.* **2015**, *139*, 55–64. [CrossRef]
24. Gautam, M.R.; Acharya, K. Streamflow trends in Nepal. *Hydrol. Sci. J.* **2012**, *57*, 344–357. [CrossRef]
25. Immerzeel, W.W.; Van Beek, L.P.H.; Konz, M.; Shrestha, A.B.; Bierkens, M.F.P. Hydrological response to climate change in a glacierized catchment in the Himalayas. *Clim. Chang.* **2012**, *110*, 721–736. [CrossRef] [PubMed]
26. Panthi, J.; Dahal, P.; Shrestha, M.L.; Aryal, S.; Krakauer, N.Y.; Pradhanang, S.M.; Lakhankar, T.; Jha, A.; Sharma, M.; Karki, R. Spatial and temporal variability of rainfall in the Gandaki River Basin of Nepal Himalaya. *Climate* **2015**, *3*, 210–226. [CrossRef]
27. Khatiwada, K.R.; Panthi, J.; Shrestha, M.L.; Nepal, S. Hydro-climatic variability in the Karnali River Basin of Nepal Himalaya. *Climate* **2016**, *4*, 17. [CrossRef]
28. Shrestha, S.; Shrestha, M.; Babel, M.S. Modelling the potential impacts of climate change on hydrology and water resources in the Indrawati River Basin, Nepal. *Environ. Earth Sci.* **2016**, *75*, 280. [CrossRef]
29. Nepal, S. Impacts of climate change on the hydrological regime of the Koshi river basin in the Himalayan region. *J. Hydro-Environ. Res.* **2016**, *10*, 76–89. [CrossRef]
30. Agarwal, A.; Babel, M.S.; Maskey, S. Analysis of future precipitation in the Koshi river basin, Nepal. *J. Hydrol.* **2014**, *513*, 422–434. [CrossRef]
31. Devkota, L.P.; Gyawali, D.R. Impacts of climate change on hydrological regime and water resources management of the Koshi River Basin, Nepal. *J. Hydrol. Reg. Stud.* **2015**, *4*, 502–515. [CrossRef]
32. Sharma, R.H.; Shakya, N.M. Hydrological changes and its impact on water resources of Bagmati watershed, Nepal. *J. Hydrol.* **2006**, *327*, 315–322. [CrossRef]
33. Babel, M.S.; Bhusal, S.P.; Wahid, S.M.; Agarwal, A. Climate change and water resources in the Bagmati River Basin, Nepal. *Theor. Appl. Climatol.* **2014**, *115*, 639–654. [CrossRef]
34. Dahal, V.; Shakya, N.M.; Bhattarai, R. Estimating the impact of climate change on water availability in Bagmati Basin, Nepal. *Environ. Process.* **2016**, *3*, 1–17. [CrossRef]
35. Neupane, R.P.; White, J.D.; Alexander, S.E. Projected hydrologic changes in monsoon-dominated Himalaya Mountain basins with changing climate and deforestation. *J. Hydrol.* **2015**, *525*, 216–230. [CrossRef]
36. DWIDP. *Hydrological Study and Data Collection of Rosi River Catchment*; A Technical Report of Department of Water Induced Disaster Prevention; Government of Nepal and Recham Consulting Pvt Ltd.: Kathmandu, Nepal, 2011.
37. Costa, A.C.; Soares, A. Trends in extreme precipitation indices derived from a daily rainfall database for the South of Portugal. *Int. J. Climatol.* **2009**, *29*, 1956–1975. [CrossRef]
38. Peterson, T.; Folland, C.; Gruza, G.; Hogg, W.; Mokssit, A.; Plummer, N. *Report on the Activities of the Working Group on Climate Change Detection and Related Rapporteurs*; World Meteorological Organization: Geneva, Switzerland, 2001.
39. Mann, H.B. Nonparametric tests against trend. *Econom. J. Econom. Soc.* **1945**, *3*, 245–259. [CrossRef]
40. Kendall, M.G. *Rank Correlation Methods*; Oxford University Press: Oxford, UK, 1948.
41. Hamed, K.H. Trend detection in hydrologic data: The Mann–Kendall trend test under the scaling hypothesis. *J. Hydrol.* **2008**, *349*, 350–363. [CrossRef]
42. Sen, P.K. Estimates of the regression coefficient based on Kendall's tau. *J. Am. Stat. Assoc.* **1968**, *63*, 1379–1389. [CrossRef]
43. Uprety, Y.; Shrestha, U.B.; Rokaya, M.B.; Shrestha, S.; Chaudhary, R.P.; Thakali, A.; Cockfield, G.; Asselin, H. Perceptions of climate change by highland communities in the Nepal Himalaya. *Clim. Dev.* **2017**, *9*, 649–661. [CrossRef]
44. Central Bureau of Statistics (CBS). *National Climate Change Impact Survey 2016*; A Statistical Report; Central Bureau of Statistics, Government of Nepal: Kathmandu, Nepal, 2017.

45. Practical Action. *Temporal and Spatial Variability of Climate Change over Nepal (1976–2005)*; Practical Action Nepal Office: Kathmandu, Nepal, 2009.
46. Merz, J.; Dangol, P.M.; Dhakal, M.P.; Dongol, B.S.; Nakarmi, G.; Weingartner, R. Rainfall-runoff events in a middle mountain catchment of Nepal. *J. Hydrol.* **2006**, *331*, 446–458. [CrossRef]
47. Olsson, O.; Gassmann, M.; Wegerich, K.; Bauer, M. Identification of the effective water availability from streamflows in the Zerafshan river basin, Central Asia. *J. Hydrol.* **2010**, *390*, 190–197. [CrossRef]
48. Liu, J.; Wang, Z.; Gong, T.; Uygen, T. Comparative analysis of hydroclimatic changes in glacier-fed rivers in the Tibet-and Bhutan-Himalayas. *Quat. Int.* **2012**, *282*, 104–112. [CrossRef]
49. Blöschl, G.; Ardoin-Bardin, S.; Bonell, M.; Dorninger, M.; Goodrich, D.; Gutknecht, D.; Matamoros, D.; Merz, B.; Shand, P.; Szolgay, J. At what scales do climate variability and land cover change impact on flooding and low flows? *Hydrol. Process. Int. J.* **2007**, *21*, 1241–1247. [CrossRef]
50. Hao, X.; Chen, Y.; Xu, C.; Li, W. Impacts of climate change and human activities on the surface runoff in the Tarim River Basin over the last fifty years. *Water Resour. Manag.* **2008**, *22*, 1159–1171. [CrossRef]
51. Ghimire, C.P.; Bruijnzeel, L.A.; Lubczynski, M.W.; Bonell, M. Rainfall interception by natural and planted forests in the Middle Mountains of Central Nepal. *J. Hydrol.* **2012**, *475*, 270–280. [CrossRef]
52. Ellison, D.; Futter, N.; Bishop, K. On the forest cover–water yield debate: From demand-to supply-side thinking. *Glob. Chang. Biol.* **2012**, *18*, 806–820. [CrossRef]
53. ICIMOD (International Centre for Integrated Mountain Development). Land cover of Nepal 2010. Available online: http://rds.icimod.org/Home/DataDetail?metadataId=9224 (accessed on 3 March 2016).

© 2018 by the authors. Licensee MDPI, Basel, Switzerland. This article is an open access article distributed under the terms and conditions of the Creative Commons Attribution (CC BY) license (http://creativecommons.org/licenses/by/4.0/).

Article

Evaluation of Gridded Multi-Satellite Precipitation Estimation (TRMM-3B42-V7) Performance in the Upper Indus Basin (UIB)

Asim Jahangir Khan [1,2,]*, Manfred Koch [1] and Karen Milena Chinchilla [1]

[1] Department of Geohydraulics and Engineering Hydrology, University of Kassel, Mönchebergstraße 19, 34125 Kassel, Germany; kochm@uni-kassel.de or manfred_kochde@yahoo.de (M.K.); chatiking@gmail.com (K.M.C.)
[2] Department of Environmental Sciences, COMSATS Institute of Information Technology, Abbottabad Campus, University Road, Tobe Camp, 22060 Abbottabad, Khyber Pakhtunkhwa, Pakistan
* Correspondence: asimjkw@gmail.com; Tel.: +49-17631674283

Received: 18 August 2018; Accepted: 5 September 2018; Published: 7 September 2018

Abstract: The present study aims to evaluate the capability of the Tropical Rainfall Measurement Mission (TRMM), Multi-satellite Precipitation Analysis (TMPA), version 7 (TRMM-3B42-V7) precipitation product to estimate appropriate precipitation rates in the Upper Indus Basin (UIB) by analyzing the dependency of the estimates' accuracies on the time scale. To that avail, various statistical analyses and comparison of Multisatellite Precipitation Analysis (TMPA) products with gauge measurements in the UIB are carried out. The dependency of the TMPA estimates' quality on the aggregation time scale is analyzed by comparisons of daily, monthly, seasonal and annual sums for the UIB. The results show considerable biases in the TMPA Tropical Rainfall Measurement Mission (TRMM) precipitation estimates for the UIB, as well as high numbers of false alarms and miss ratios. The correlation of the TMPA estimates with ground-based gauge data increases considerably and almost in a linear fashion with increasing temporal aggregation, i.e., time scale. There is a predominant trend of underestimation of the TRMM product across the UIB at most of the gauge stations, i.e., TRMM-estimated rainfall is generally lower than the gauge-measured rainfall. For the seasonal aggregates, the bias is mostly positive for the summer but predominantly negative for the winter season, thereby showing a slight overestimation of the precipitation in summer and underestimation in winter. The results of the study suggest that, in spite of these discrepancies between TMPA estimates and gauge data, the use of the former in hydrological watershed modeling undertaken by the authors may be a valuable alternative in data-scarce regions like the UIB, but still must be taken with a grain of salt.

Keywords: Precipitation; Tropical Rainfall Measurement Mission (TRMM); Multi-Satellite Precipitation Analysis (TMPA); Upper Indus Basin (UIB)

1. Introduction

The continued improvements in computation capabilities and the subsequent increase in the development of spatially explicit and distributed models for expressing environmental phenomena have necessitated the provision of more intensive and improved data for environmental variables both in space and time. Two major issues, particularly in hydro-meteorological studies, are the possible sparsity of data sampling points (gauge stations), and the discontinuities in the data and in the quality of the temporal records. These issues are more frequent in mountainous regions with high altitudes, because they are immensely challenging environments for measurements of precipitation through remote sensing or traditional ground-based methods due to the difficult topography and the highly-variable weather and climatic conditions [1,2]. These factors have proved to be the main hurdles due to which many developing countries are unable to achieve consistent spatial and temporal coverage for ground-based precipitation measurements [2,3], therefore making it difficult for them to achieve an effective spatial coverage of rainfall [4,5]. The consequent lack of good quality precipitation data is thus a big hurdle for properly assessing impacts of climate change on water resources in these regions [1].

As data with an acceptable gridded resolution of daily climatic variables are critical for hydrological and water resources modeling [6,7], managing the gaps in the data appropriately is then the first stage of most climatological, environmental, and hydrological studies [2]. This step is also necessary to improve the spatial resolution for sparse gauge station data sets before using them as an input for spatially-distributed rainfall-runoff models, because the gauge-based interpolation methods, commonly used in hydrologic models, usually do not cover the spatial heterogeneity of the variability of climatic variables in the catchment. These errors in the interpolated data field then have the potential to significantly bias model calibrations and water balance calculations [6].

Fortunately, continued scientific developments are also providing new prospects for addressing these issues. For example, the advancements in gathering and deriving climate data through satellite remote sensing could provide a possible opportunity to address some of the issues with regard to the spatial coverage of climate data. That is why the use of satellite-based precipitation products individually or in combination with land-based gauge data has been increasingly recognized as a very promising alternative to address the aforementioned problems [5]. Such precipitation products have proven to be of great value, especially in developing countries with remote and high-altitude locations where conventional rain gauge or weather data are of bad quality or have low coverage [8].

There are numerous satellite-based precipitation products currently available, with varying degrees of accuracy. These include the Climate Prediction Center (CPC) morphing algorithm (CMORPH) [9], Global Satellite Mapping of Precipitation [10–12], Naval Research Laboratory Global Blended Statistical Precipitation Analysis [13], the Tropical Rainfall Measuring Mission (TRMM), Multisatellite Precipitation Analysis (TMPA) [14,15], and a few others. Since their inception, most of these gridded datasets have been evaluated for their suitability and usability for specific regions or intended uses. In general, such investigations are less frequently carried out for mountainous regions and even less so for our study area, the Hindukush, Karakurm, and Himalays (HKH) region. In the HKH and the Upper Indus Basin (UIB) region, most of the reported works related to the evaluation of gridded precipitation products are suggestive of considerable biases in the gridded products in comparison to the gauge records [16–19].

Additionally, the quality, coverage, and representativeness of the available observed gauge records in the HKH have also been questioned and have sometimes been regarded as having considerable underestimations of regional precipitation amounts [20–22], especially at higher altitudes [23]. The spatial distribution of estimated real precipitation by Khan and Koch (unpublished) over the study area is given in Figure A1, while the vertical meteorological and cryspheric regimes in the UIB (modified from Hewitt 2007) are given in Figure A2.

While in most cases these gridded global precipitation data sets are some interpolated version of point measurements (most often through geo-statistical procedures), they may only be useful for regions where dense network of rain gauges are available, because otherwise, in the absence of a dense enough networks or over regions of complex topographies, the interpolated precipitation present a very generalized restitution which is not able to reflect properly the prevalent orographic, surface, or atmospheric processes [19].

In comparison with the sparse gauge observations or gridded data products based on them, satellite-based precipitation products such as Tropical Rainfall Measurement Mission (TRMM) Multi-satellite Precipitation Analysis (TMPA), version 7 (TRMM-3B42-V7) have an inherent advantage due to their higher spatial coverage. However, they also have certain limitations because they are indirect estimates of rainfall, which depend on the cloud height and the properties of the cloud's surfaces (IR-algorithms) and on the integrated sparse and multi-source hydro-meteorological content (passive microwave algorithms) [15,24,25]. Therefore, before such satellite-based data can be used with confidence, it is important to evaluate their accuracy or error characteristics by comparing them with data from ground-based observations.

Based on these considerations, the current study aims at assessing the skill of the TRMM precipitation dataset in matching the magnitudes and occurrences at different temporal scales at all the points of the observational network available in order to evaluate its further processing and correction requirements or suitability for subsequent use in distributed hydrological modeling.

2. Materials and Methods

2.1. Study Area and Data

2.1.1. Study Area: The Upper Indus River Basin (UIB)

The Indus River, one of the largest rivers in Asia with a total length of about 2880 km, has a drainage area of about 912,000 km^2 which extends across portions of India, China, Pakistan, and Afghanistan. The portion of the Indus that comprises the Upper Indus River Basin (UIB), with a logical lower boundary at Tarbela Dam (Figure 1), is about 1125 km long and drains an area of about 170,000 km^2 [26].

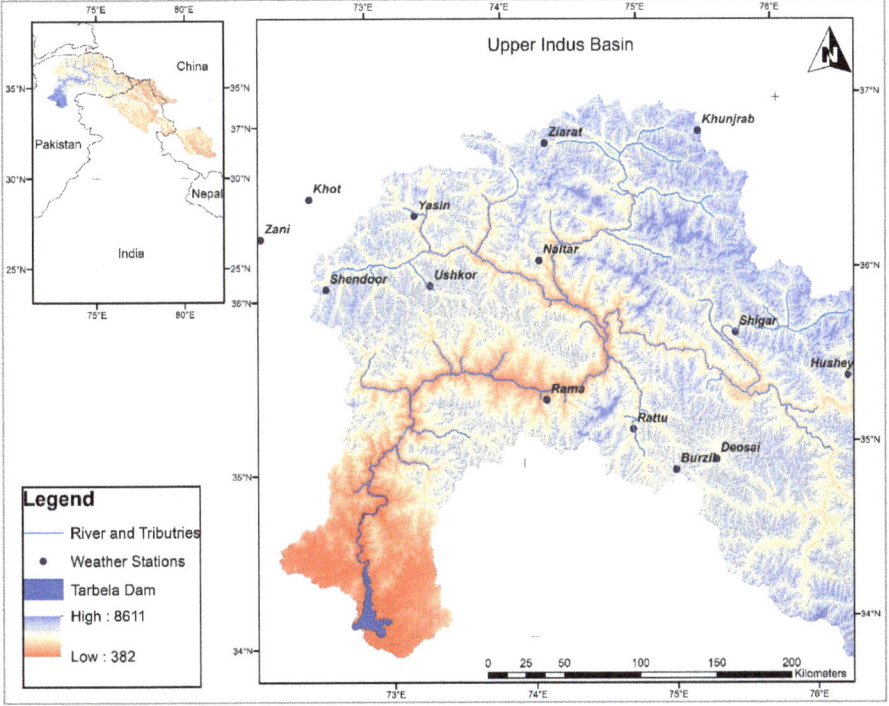

Figure 1. Upper Indus Basin with locations of hydro-climatological stations.

Being a high-mountain region, the UIB contains the largest area of perennial glacial ice cover (22,000 km^2) outside the polar regions of the earth, and which extends even further during the winter season [27]. The altitude within the UIB ranges from as low as 455 m to a high of 8611 m and, as a result, the climate varies greatly within the basin [28].

The summer monsoon has no significant effect on the basin, as almost 90% of its area lies in the rain shadow of the Himalayan belt [20,29]. Except for the south-facing foothills, the intrusion of the Indian Ocean monsoon is limited by the mountains so that its influence weakens northwestward. Subsequently, the climatic controls in the UIB are quite different from that in the Himalayas on the eastern side. In fact, over the extent of the UIB, most of the annual precipitation originates in the west and falls in winter and spring, whereas occasional rains are brought by the monsoonal incursions to the trans-Himalayan areas. Even during the summer months, the trans-Himalayan areas do not obtain all their precipitation from the monsoons [30–33].

Climatic variables are usually strongly influenced by topographic altitude. Several studies have reported that precipitation in the HKH region exhibits large changes over short distances and has a considerable vertical gradient [30,31,34–37]. Thus, the northern valley floors of the UIB are arid, with annual precipitation of only 100–200 mm, but these totals increase with elevation and reach up to 600 mm at 4400 m height and lead to annual glacier accumulation rates of up to 1500 to 2000 mm at 5500 m altitude, according to some glaciological studies [30]. The average snow cover area in the Upper Indus River Basin fluctuates between 10% and 70%. Snow cover in the area is at a maximum of 70–80% in the winter snow accumulation period (December to February) and at a minimum of 10–15% in the summer snow melt period (June to September) [28]. Stream flow is generated by the combination of the storm runoff in the lower parts of the upper Indus basin and the snow and glacier runoff from the higher parts of the UIB [26,38].

2.1.2. Data

TMPA Data (TRMM-3B42-V7)

In this study the TRMM-3B42-V7 precipitation product is used. This product is basically a calibration-based combination scheme for precipitation estimates from multiple satellites and space-borne sensors, including infrared, microwave, radar data, and gauge measurements. Though the dataset has very good spatio-temporal resolution (0.25° × 0.25° grid, 3-hourly) and good global coverage (latitude band 50° N to 50° S) and has data available from 1998 to the recent past [1,15]. It also has some uncertainties, because the inputs on which it is based are indirect estimates of rainfall, depending on the cloud height and the properties of the cloud surface (IR algorithms) and on the integrated sparse and multi-source hydro-meteorological content (passive microwave algorithms) [14,15,25].

During the current study, 3-hourly data from 1 January 1998 to 31 December 2008 were summed to daily accumulated precipitation for each of the 0.25° × 0.25° grid boxes (which have a gauge station), and evaluated for matches with the corresponding gauge station's observed daily accumulated precipitation. As the observational network is scant, no TRMM grid box included more than one in situ gauge station.

Observed Ground-Based Data

In the HKH region of Pakistan, observed in situ data are limited and operated by different organizations, mainly the Pakistan Meteorological Department (PMD) and Water and Power Development Authority (WAPDA). The stations operated by PMD (Figure 1) have daily-time-step climate data available for longer periods (1947 to date), but with huge gaps and missing data in the records and with only monthly data available freely for research purposes. Furthermore, all the PMD stations are valley-based, at elevations below 3000 m.a.s.l. in altitude and, therefore hardly represent the frequency and amount of precipitation in the high-altitude areas. The climate stations, operated by the WAPDA, are fairly new and have considerably consistent data over the time period, coinciding with the TRMM product. These gauge stations are distributed almost evenly across the UIB inside Pakistan and cover a wide range of elevations.

During the current study, daily precipitation records of 14 meteorological stations operated by WAPDA are utilized for the evaluation of the TRMM estimates. Their geographical attributes are given in Table 1. The evaluation is limited to the duration of 1998 to 2008, as the observed precipitation data could not be acquired for the period beyond 2008.

Table 1. Geographical attributes of the precipitation gauge network.

Description	No.	Station	Latitude (o)	Longitude (o)	Altitude (m)
	1	Burzil	34.91	75.90	4030
	2	Deosai	35.09	75.54	4149
	3	Hushey	35.42	76.37	3075
	4	Khot	36.52	72.58	3505
High Altitude	5	Khunjrab	36.84	75.42	4440
(2367–4440 m.a.s.l.)	6	Naltar	36.17	74.18	2898
stations operated by	7	Rama	35.36	74.81	3179
Water and Power	8	Rattu	35.15	74.8	2718
Development Authority	9	Shendoor	36.09	72.55	3712
(WAPDA), Pakistan	10	Shigar	35.63	75.53	2367
	11	Ushkor	36.05	73.39	3051
	12	Yasin	36.45	73.3	3350
	13	Zani	36.33	72.17	3895
	14	Ziarat	36.77	74.46	3020

2.2. Methods

The quantitative comparison of the TRMM estimates with the ground rain-gauge station observations is done by employing various widely used statistical indicators. These include the correlation coefficient (r), the mean relative bias error (rBIAS), the mean bias error (MBE), the mean absolute error (MAE), and the root mean square error (RMSE), defined in the following equations:

$$r = \frac{\sum_{i=1}^{n}(T_i - \overline{T})(G_i - \overline{G})}{\sqrt{\sum_{i=1}^{n}(T_i - \overline{T})^2}\sqrt{\sum_{i=1}^{n}(G_i - \overline{G})^2}} \tag{1}$$

$$rBIAS = \frac{1}{n}\sum_{i=1}^{n}\left(\frac{T_i - G_i}{G_i}\right) \tag{2}$$

$$MAE = \frac{1}{n}\sum_{i=1}^{n}|T_i - G_i| \tag{3}$$

$$MBE = \frac{1}{n}\sum_{i=1}^{n}(T_i - G_i) \tag{4}$$

$$RMSE = \sqrt{\frac{1}{n}\sum_{i=1}^{n}(T_i - G_i)^2} \tag{5}$$

where n is the number of samples, T_i refers to satellite-based precipitation, G_i is gauge-based precipitation, and $\overline{T}$ and $\overline{G}$ are the corresponding means. Among these statistical indices, r shows the degree of linear correlation between TRMM precipitation estimates and gauge observations. MBE, MAE, and rBIAS are used to assess the systematic bias, i.e., the deviation of the satellite precipitation from the gauge observations, and the RMSE gives the magnitude of the average error in relative terms.

In addition, evaluations were also made for the daily TRMM estimates and gauge data over the Indus river basin, based on a 2 × 2 contingency table (Table 2), by detecting rain events (Hits), no events (Correct Negative), Misses by the TRMM, and False Alarms by the TRMM. More specifically, we used a threshold of 0.3 mm/d to differentiate precipitation and no precipitation events, since lower precipitation values may be the result of noise, as indicated by [31,39] etc.

Table 2. Contingency Table 2 × 2. TRMM: Tropical Rainfall Measurement Mission.

		Observed Values (Gauge Data)		Total
		YES	NO	
Estimated Values (TRMM estimates)	YES	-a -Hits	-b -False Alarms	Total—Yes Estimated
	NO	-c -Misses	-d -Correct Negative	Total—No Estimated
Total		Total—Yes Observed	Total—No Observed	Total $a + b + c + d$

Based on these four indicators, with orders as shown in the table, several categorical statistical indices are derived, including accuracy (*Ac*), bias score or frequency bias index (*FBI*), probability of detection (*POD*), false alarm ratio (*FAR*), critical success index (*CSI*), and true skill statistics (*TSS*) [40,41], defined in the following equations:

$$Ac = \frac{a+d}{Total} \quad (6)$$

$$FBI = \frac{a+b}{a+c} \quad (7)$$

$$POD = \frac{a}{a+c} \quad (8)$$

$$FAR = \frac{b}{a+b} \quad (9)$$

$$CSI = \frac{a}{a+b+c} \quad (10)$$

$$TSS = \frac{a}{a+b} - \frac{b}{c+d} = \frac{ad-bc}{(a+b)(c+d)} \quad (11)$$

where *a* represents the number of rainfall events that have been successfully estimated by TRMM data (hits), *b* is the number of events incorrectly predicted as rain events by TRMM (false alarms), *c* is the number actual events that are missed by TRMM (Misses), and *d* is the number of dry days or no-rainfall events identified successfully by the TRMM dataset. For each day, depending on how the estimated and observed precipitation behave, any event above the given threshold (0.3 mm) is scored either as a Hit, Miss, False Alarm or Correct Negative, so that the rainfall is a Hit if both TRMM and observed precipitation reach the threshold; False Alarm if only the TRMM estimate reaches the threshold; Miss if only the observed precipitation reaches it; and Correct Negative if both are below the threshold. The number of Hits, False Alarms, Misses and Correct Negatives are used in Equations (5)–(10) to calculate the above mentioned statistical indices.

Each of these indices provides a specific information on the two data sets compared. Thus, *Ac* indicates the fraction of estimates which is correct (range: 0 to 1, perfect score: 1); *FBI* indicates whether the estimated dataset has a tendency to underestimate (*FBI* < 1) or to overestimate (*FBI* > 1) rain events; *POD* quantifies the fraction of rain occurrences that is estimated correctly (range: 0 to 1, perfect score: 1); *FAR* measures the fraction of false alarms in the satellite rain estimates (range from 0 to 1, perfect score: 0); and *CSI* measures the fraction of estimated events that are correctly predicted (range from 0 to 1, perfect score: 1). Unlike all the aforementioned indices, *TSS* does not depend on the frequency of the climatological event and uses all elements in the contingency table (Table 2). Thus, *TSS* provides a measure of the accuracy of the estimates in terms of the probability of correct detection of events or no events. In this case the range is from −1 to 1, with a perfect score being 1, with 0 showing no skills, and a negative score signifying that the estimates are worse than a random forecast.

3. Results and Discussion

The assessment of the reliability of the TRMM estimates and their comparisons with the rain data from gauge station presented in this section has been done by three different methodologies: (1) a *statistical analysis*, based on *r*, *rBIAS*, *MBE*, *MAE*, and *RMSE* for daily, monthly, annual and seasonal data aggregates; (2) *categorical statistics* of daily data by computing *Ac*, *FBI*, *POD*, *FAR*, *CSI*, and *TSS*; and (3) a *visual comparison* for monthly, annual, and seasonal data.

3.1. Statistical Analysis for Daily, Monthly, Annual, and Seasonal Agreggates

The results of the TRMM-assessment based on the statistical measures *r*, *rBIAS*, *MBE*, *MAE*, and *RMSE* are given for daily data aggregation in Table 3, for monthly and annual data aggregation in Table 4, and for seasonal aggregation in Table 5. The summer seasons include the months of April, May, June, July, August, and September, while the remaining 6 months, which are October, November, December, January, February, and March, are aggregated to represent the winter season.

It is evident from a first glance at the two tables that the TRMM performs overall rather poorly for estimating the observed rain amounts for the study region at all temporal scales, as the average *r* values are only 0.16, 0.22, 0.22 and 0.20 for monthly, annual, and seasonal (summer and winter) aggregation, respectively. Further specific results are discussed in the subsequent sub-sections.

3.1.1. Skill Statistics for TRMM Precipitation Estimates (Daily Aggregates)

The daily aggregates of the TRMM precipitation estimates show poor skill in matching observed precipitation, with an average *r* of only 0.16 (Table 3). The comparison of the observed and TRMM daily rainfall data indicates highly variable *MAE* values across the UIB, with a range ≥ 23 mm/day (Table 3). Values of *MAE* are high throughout most of compared locations in the UIB, with $MAE \leq 13$ mm for all stations across the UIB. The northwestern parts of the UIB show the highest and most variable *MAE*.

Table 3. Statistical analysis for daily aggregates. UIB: Upper Indus Basin; *r*: correlation coefficient; *rBIAS*: mean relative bias error; *MBE*: mean bias error; *MAE*: mean absolute error; *RMSE*: root mean square error.

Sub-Basin	Station	Daily				
		r	*rBIAS*	*MBE* (mm)	*MAE* (mm)	*RMSE* (mm)
Southern UIB	Burzil	0.22	−0.42	−11.77	16.20	20.76
	Deosai	0.10	0.99	4.41	14.53	26.33
	Rama	0.23	−0.22	−16.20	18.31	27.44
	Rattu	0.14	0.69	−7.90	18.12	29.24
Eastern UIB	Shigar	0.08	1.31	−3.99	10.24	17.81
	Hushey	0.14	−0.07	−5.73	10.79	14.73
Northwestern UIB	Khot	0.19	0.70	0.49	4.93	8.08
	Naltar	0.25	0.24	−11.94	15.39	21.10
	Shendoor	0.16	1.48	1.22	9.46	15.67
	Ushkor	0.21	0.70	−0.51	8.16	12.74
	Yasin	0.10	5.27	24.24	28.68	46.57
	Zani	0.13	−0.14	−15.23	19.16	26.96
Northern UIB	Khunjrab	0.15	−0.27	−5.50	10.52	15.40
	Ziarat	0.14	0.71	−0.94	6.05	9.48
Basin average		0.16	0.78	−3.53	13.61	20.88
Maximum		0.25	5.27	24.24	28.68	46.57
Minimum		0.08	−0.42	−16.20	4.93	8.08

Based on the values of the *MBE* in Table 3, one can notice that the TRMM data have huge underestimations across most of the UIB (average *MBE* of −3.53 mm), while they show at the same time a distinct spatial pattern across the study area, with a clear underestimation of the TRMM estimates for all the studied locations in the eastern and northern UIB, as well as for the southern UIB at all locations except one. On the other hand, the stations located in the northwestern UIB experience a mixed trend, with TRMM data indicating a moderate to high under- or overestimation at half (three) of the locations each.

The mean relative bias (*rBIAS*) at the different gauge locations also follows a similar pattern, with huge variations and range from a negative −0.42 to a high of +5.27 at station "Yasin". The TRMM precipitation estimate is thus more than 5 times the gauge-based observed value, which means that there is tremendous overestimation of the former.

For the relative bias *rBias*, the TRMM estimates show underestimation only at five locations, unlike for the mean bias *MBE* discussed before, where 10 out of the 14 locations display underestimations. After a more detailed examination of both the TRMM- and observed time series, it was found that, at some locations, very large number of light precipitation events were generally overestimated, while a smaller number of heavy precipitation events were underestimated by the TRMM estimates. This behavior may have resulted in the overall average positive *rBIAS* at most of the locations, discussed earlier, while in reality the overall mean bias *MBE* of the TRMM estimates is negative (underestimation) at more than two-thirds of the locations.

The *RMSE* values for the daily time series are also very high and show large variations, ranging from as low as 8.08 mm/day to as high as 46.57 mm/day, with an average basin-wide value of $RSME = 20.88$ mm/day.

These results are in general agreement with previous studies [23,42], as most of them have reported the TRMM product to underestimate the gauge-based rainfall amounts over the HKH region in general and, even more so over the western parts of HKH.

3.1.2. Skill Statistics for TRMM Precipitation Estimates (Monthly and Annual Aggregates)

The skill statistics of the monthly and annually aggregated TRMM precipitation estimates listed in Table 4 shows overall also poor skill in matching the observed ground-based precipitation, but with considerably improved values for the Pearson correlation coefficient *r* for all the studied locations, namely, with *r* values of 0. 61 and 0.57 for the basin average rainfall for monthly and annual aggregates, respectively.

Table 4. Statistical analysis based on monthly and annual data aggregation.

Sub-Basin	Station	Monthly					Annual				
		r	rBIAS	MBE (mm)	MAE (mm)	RMSE (mm)	r	rBIAS	MBE (mm)	MAE (mm)	RMSE (mm)
Southern UIB	Burzil	0.55	−0.28	−56.80	58.73	112.8	0.43	−1.05	−391.5	391.5	407.7
	Deosai	0.22	0.17	21.65	43.62	78.95	−0.37	0.24	146.6	201.4	234.6
	Rama	0.54	−0.36	−78.26	79.54	165.9	0.78	−2.05	−538.5	538.5	574.8
	Rattu	0.20	−0.20	−39.31	61.32	119.10	0.30	−0.58	−264.9	274.1	353.8
Eastern UIB	Shigar	0.02	−0.21	−19.42	36.56	78.26	−0.11	−0.62	−133.2	179.7	249.8
	Hushey	0.09	−0.24	−29.19	39.72	101.9	−0.15	−0.79	−193.2	230.3	339.6
Northwestern UIB	Khot	0.51	−0.21	−26.52	39.47	74.14	0.29	−0.65	−182.0	242.7	250.9
	Naltar	0.60	−0.31	−58.62	60.17	120.6	0.12	−1.32	−395.1	395.1	416.3
	Shendoor	0.42	0.08	6.53	22.08	37.34	0.54	0.13	40.4	66.4	99.1
	Ushkor	0.52	−0.03	−1.82	19.19	39.41	0.62	−0.05	−14.7	79.1	110.5
	Yasin	0.21	1.36	118.3	126.7	241.2	0.10	0.72	806.0	806.0	827.5
	Zani	0.49	−0.34	−75.05	77.98	154.91	0.52	−1.69	−508.3	508.3	532.6
Northern UIB	Khunjrab	0.39	0.05	2.26	13.13	23.40	−0.28	0.09	18.3	52.0	70.0
	Ziarat	0.38	−0.07	−5.39	15.23	32.62	0.55	−0.15	−30.6	73.4	104.7
Basin average		0.61	−0.23	−9.09	14.15	20.98	0.57	−0.24	−117.3	117.3	134.1
Maximum		0.60	1.36	118.3	126.7	241.2	0.78	0.72	806.0	806.0	827.5
Minimum		0.02	−0.36	−78.26	13.13	23.40	−0.37	−2.05	−538.50	51.96	69.96

The *MAE* for these long-term aggregated precipitation data is also highly variable across the UIB, ranging from 13.13 mm/month to 126.68 mm/month for the monthly aggregates and from

−538.5 mm/year to 806.0 mm/year for annual aggregates. The values for *MAE* are generally high throughout most of the locations in the UIB, with a basin-wide average of 14.15 mm/month and 117.3 mm/year for monthly and annual aggregated rainfall, respectively. The spatial pattern of the errors observed in both cases as well as the predominant underestimations observed at most locations is similar to that observed for the daily aggregates above. Thus the northwestern parts of the UIB show the highest and the most variable *MAE*.

The TRMM estimates are also significantly underestimated across most of the UIB, with and average basin-wide *MBE* of −9.09 mm/month and −117.3 mm/year for monthly and annual aggregated rainfall, respectively. The eastern part of UIB shows a distinct underestimation of the TRMM rainfall for all stations there and this underestimation is even higher in the southern UIB, where three out of the four locations experience this effect, while at one location (Deosai), an overestimation of 21.65 mm/month for the monthly aggregate is observed. The stations located in the northwestern UIB have a mixed trend, with moderate-to- high underestimations at four locations and an opposite behavior in the remaining two, and this holds for both monthly and annual aggregates. For the northern UIB, for the *MBE* of the two locations evaluated there have also been mixed results, with one station (Khunjrab) indicating a slight overestimation with *MBE* of 0.05 mm/month and 18.3 mm/year, and the other (Ziarat) an underestimation with a negative *MBE* of −0.07 mm/month and −30.6 mm/year for monthly and annual aggregated rainfall, respectively.

3.1.3. Skill Statistics for TRMM Precipitation Estimates (Seasonal Aggregates)

The various statistical indices of the two seasonally (summer and winter) aggregated TRMM estimates are shown in Table 5.

Table 5. Statistical analysis for summer and winter season data aggregation.

Sub-Basin	Station	Summer Season					Winter Season				
		r	rBIAS	MBE (mm)	MAE (mm)	RMSE (mm)	r	rBIAS	MBE (mm)	MAE (mm)	RMSE (mm)
Southern UIB	Burzil	0.35	−0.48	−190.8	190.8	201.5	0.05	−0.50	−200.7	200.7	226.0
	Deosai	−0.42	0.27	68.1	101.0	123.5	−0.14	0.31	78.6	116.9	139.6
	Ramma	0.68	−0.54	−198.7	198.7	213.9	0.60	−0.92	−339.9	339.9	371.4
	Rattu	0.49	0.09	25.8	101.0	127.9	−0.01	−1.00	−290.8	290.8	357.1
Eastern UIB	Shigar	−0.09	−0.22	−36.9	106.0	156.6	−0.10	−0.57	−96.3	105.4	127.9
	Hushey	0.22	−0.35	−83.2	103.7	156.3	−0.55	−0.46	−109.9	132.2	189.3
Northwestern UIB	Khot	−0.11	−0.03	−4.1	38.1	48.3	0.33	0.16	22.2	29.2	34.8
	Naltar	0.48	−0.50	−203.8	203.8	220.6	0.17	−0.47	−191.6	191.6	205.9
	Shendoor	0.57	0.42	56.1	56.7	79.6	−0.02	−0.12	−15.9	61.1	65.6
	Ushkor	0.49	0.17	31.0	71.9	84.1	0.65	−0.25	−45.9	55.3	85.7
	Yasin	−0.35	2.94	593.6	593.6	615.2	0.22	1.05	212.5	212.5	244.1
	Zani	0.52	−0.66	−268.8	268.8	301.3	0.57	−0.59	−239.8	239.8	256.1
Northern UIB	Khunjrab	0.09	−0.21	−41.3	91.8	109.2	0.56	−0.72	−141.1	157.5	166.9
	Ziarat	0.56	−0.02	−2.5	36.2	47.4	0.54	−0.20	−28.4	44.8	65.3
Basin average		0.60	0.00	−5.18	30.1	38.7	0.36	−0.4	−96.51	96.5	109.1
Maximum		0.68	2.94	593.6	593.6	615.2	0.65	1.05	212.5	339.9	371.4
Minimum		−0.42	−0.66	−268.8	36.2	47.4	−0.55	−1.00	−339.9	29.2	34.8

One may notice from the table that the various skill indices have comparable trends to those of the monthly- or annually aggregated TRMM series, in terms of magnitude, but show a different pattern for the two seasons. For example, for summer, the TRMM estimates show positive *rBIAS* for a few locations where the monthly and annual aggregates earlier indicated a negative one (i.e., stations Rattu, Ushkor). In contrast, for the winter season predominantly negative *rBIAS* values are obtained, similar to those of the monthly and annual aggregations.

The overall range of the *MBE* for the stations evaluated varies from −268.8 mm to 593.6 mm for the summer season and from −339.9 mm to 212.5 mm for the winter, with an average *MBE* across the UIB of −5.18 and −96.51 mm for summer and winter, respectively. These comparatively lower

MBE values for the summer season are indicative of a situation where the under- or overestimation occurring in the different months of the season, cancel each other partly out.

The correlations coefficient r range from -0.42 to 0.68 and -0.55 to 0.65 for the summer and winter seasons, respectively, with the average r for the UIB having better values for the summer- ($r = 0.6$) than for the winter ($r = 0.36$) season, most likely due to the fact that the precipitation is much higher in summer than in winter.

3.2. Categorical Statistics

The results of the TRMM-precipitation analysis for the six categorical indices described in Section 2.2 are listed in Table 6. Basically these indices show how the TRMM data match the ground-based gauge data at daily time scales.

Table 6. Categorical statistics for daily TRMM estimate and gauge rain data. *Ac*: accuracy; *FBI*: frequency bias index; *POD*: probability of detection; *FAR*: false alarm ratio; *CSI*: critical success index; *TSS*: true skill statistics.

Sub-Basin	Station	Ac	FBI	POD	FAR	CSI	TSS
Southern UIB	Burzil	0.57	0.75	0.42	0.45	0.31	0.12
	Deosai	0.57	1.01	0.61	0.39	0.44	0.14
	Ramma	0.55	1.27	0.50	0.61	0.28	0.08
	Rattu	0.51	1.48	0.56	0.62	0.29	0.04
Eastern UIB	Shigar	0.60	1.30	0.41	0.68	0.22	0.08
	Hushey	0.57	0.83	0.40	0.52	0.28	0.10
Northwestern UIB	Khot	0.65	1.34	0.57	0.58	0.32	0.25
	Naltar	0.62	0.86	0.40	0.53	0.28	0.14
	Shendoor	0.56	1.16	0.40	0.65	0.23	0.04
	Ushkor	0.61	1.07	0.42	0.60	0.26	0.12
	Yasin	0.67	1.03	0.44	0.58	0.27	0.20
	Zani	0.55	0.86	0.35	0.59	0.24	0.03
Northern UIB	Khunjrab	0.55	1.04	0.43	0.58	0.27	0.06
	Ziarat	0.60	0.73	0.35	0.52	0.25	0.10
Basin average		0.58	1.05	0.45	0.56	0.28	0.11
Maximum		0.67	1.48	0.61	0.68	0.44	0.25
Minimum		0.51	0.73	0.35	0.39	0.22	0.03

Thus, the values for the first index, accuracy (*Ac*) are well above 0.50 for all stations, with an average of 0.58.

The frequency bias index *FBI* has neither very high positive nor negative values, but varies on both sides, with nine stations showing overestimation, and the remaining five underestimation. The basin-wide average *FBI* is 1.05, which indicates a slight overestimation of the TRMM rainfall, vindicating the results of the general statistical skill analysis of the previous sections.

The other categorical indices (see Equations (8)–(11)) do not show very good results either. Thus, for most of the stations the values of the probability of detection (*POD*) is below 0.5, with only four stations having values above it. The false alarm ratios (*FAR*) for all stations but one are generally too high, with a basin-wide average of *FAR* = 0.56. In the same way, both the CSI and the TSS values are also not very promising, as only three stations have values above 0.30 for the former and only one station a value of about 0.20 for the latter.

Thus, overall, these results of the categorical statistics indicate that the TRMM rainfall estimates in the UIB do not have a good match with the ground-based gauge data and, therefore, should only be used after some corrections and adjustments have been made.

3.3. Visual Comparison

For visual comparison, monthly, annually, and seasonally aggregated time series of the TRMM rainfall estimates and of the various gauge stations are plotted.

Figures 2 and 3 show these time series plots for two stations, Yasin and Khunjrab, respectively. One may notice from these plots that for station Yasin (Figure 2) huge biases and errors at all three time scales considered are obtained, whereas for station Khunjrab a better match, especially at the annual and seasonal aggregations, is obtained. The corresponding plots for the others stations analyzed reveal a somewhat similar pattern.

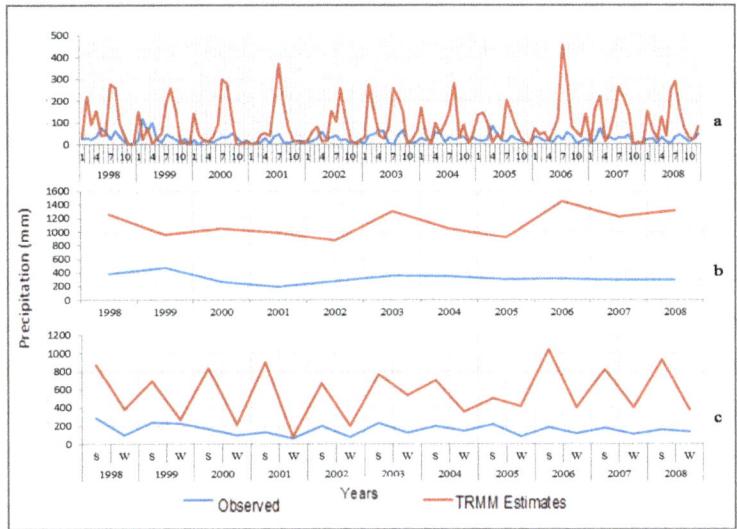

Figure 2. Time series of TRMM estimates and gauge data for rainfall totals at Yasin station; (**a**) monthly, (**b**) annual, and (**c**) seasonal (S = summer, W = winter).

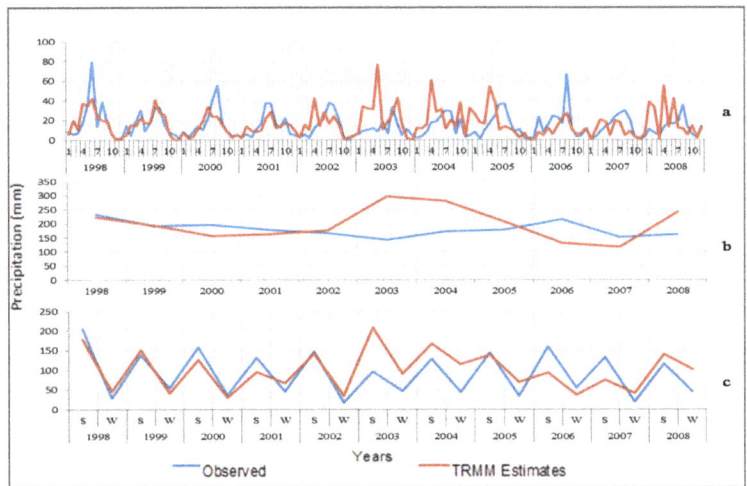

Figure 3. Time series of TRMM estimates and observed gauge data for mean rainfall totals at Khunjrab station; (**a**) monthly, (**b**) annual, and (**c**) seasonal (S = summer, W = winter).

The monthly aggregated TRMM- and gauge observed rainfalls averaged over all stations and the full length of period considered (1998–2008), are plotted in Figure 4. The figure also has a demarcation of the seasons. From the figure a systematic underestimation of the TRMM rainfall in the winter months and a mix of under and overestimation in the summer months can clearly be seen, corroborating to a large extent the statistical results of the previous sections.

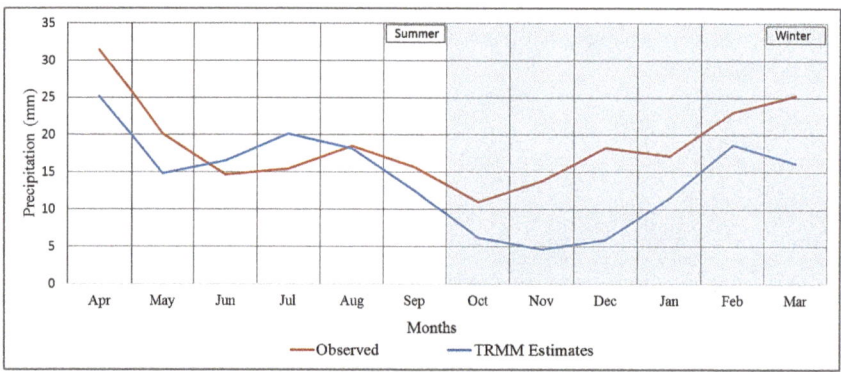

Figure 4. Average TRMM estimates and gauge data for mean monthly rainfall for all stations with seasonal demarcation.

Finally, the monthly, annually, and seasonally aggregated time series of both the averaged TRMM-rainfall estimates and of the observed gauge data are plotted in Figure 5. Though there is almost a persistent underestimation of the TRMM estimates, the peaks and troughs of both series follow, in most instances, similar patterns.

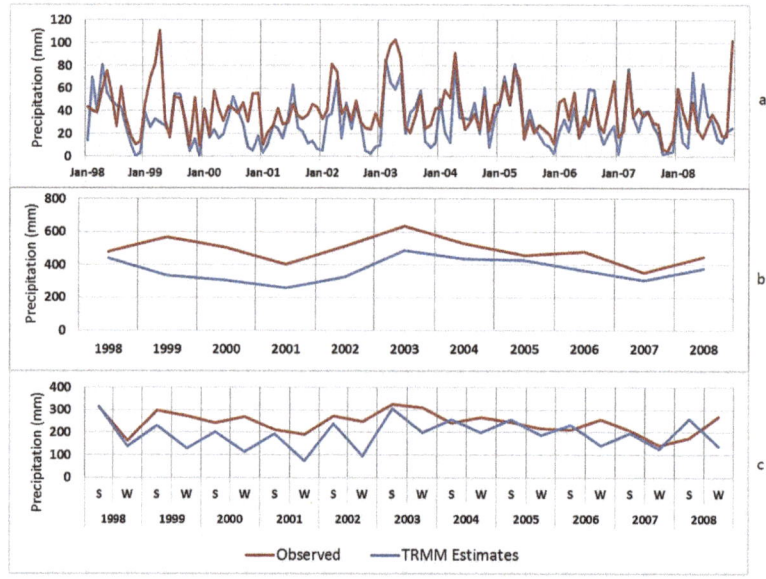

Figure 5. Time series of average TRMM estimates and observed gauge data for mean rainfall totals over the study area, for all the gauge stations; (**a**) monthly, (**b**) annual, and (**c**) seasonal (S = summer, W = winter).

4. Discussion and Conclusions

In this study, data from a TMPA product—TRMM 3B42 V7—for the Upper Indus Basin (UIB), Pakistan over the period 1998–2008 has been assessed and evaluated on a point-to-point basis by comparison with rain gauge data from 14 stations. These assessments have been performed at monthly, seasonal, and annual aggregation scales. The results indicate that the TMPA product has considerable errors in estimating the rainfall amounts at the various gauge stations throughout the study area and throughout the total time period studied. There is a predominant trend of underestimation of the TRMM product across the UIB at most of the gauge stations; TRMM-estimated rainfall is generally lower than the gauge-measured rainfall. The seasonal TRMM rainfall, though, shows a specific pattern, with the summer rainfall slightly overestimated, and winter rainfall predominantly underestimated at almost all locations and all aggregation time scales.

These results conform overall with those of previous studies, which, in most cases, suggest that neither the sparsely observed station data and gridded data products based on them, nor the sensors-based data fully represent the precipitation regime of the region [42], with strong non-representation or underestimation [16] of regional precipitation amounts, especially for higher altitudes by [20,22,42]. In fact, the in situ meteorological observations in the UIB are sparse and mostly taken at valley-based stations. These data provide low spatial coverage and are scant for higher altitudes. Furthermore, the complex orography of the UIB region also affects the amounts, spatial patterns and seasonality of the precipitation. Additionally, most of the authors [20–23,42] indicated that the observation network across the UIB also shows underestimation of precipitation amounts, with an average of around 166%, reaching in excess of 300% over some parts of the basin [23]. This means that the TRMM product may even be underestimating the true areal precipitation by a much greater margin, as the true areal precipitation is estimated to be much higher [23] than the gauge observation records.

The comparison of any gridded or sensor-based datasets with observed precipitation may not be taken, therefore, as a conclusive evidence for declaring the evaluated data is unappropriated in terms of usability, but rather shows the degree to which these data sets match the magnitudes or occurrences of the observed precipitation, which by no means is perfect, and a better match may also indicate that the evaluated data have tendencies to underestimate the real areal precipitation over the UIB. Furthermore, the spatial resolution of the TRMM product ($0.25° \times 0.25°$), may also pose limitations, especially for distributed hydrological modeling and investigations [23,43], as at this resolution, the orographic influences on the precipitation regime cannot be mapped, while the hydrological models usually require precipitation data at a much finer scale.

The main conclusion to be drawn from this study may be then resumed as follows:

(1) The TRMM-3B42-V7 product has an overall poor agreement with the observed rainfall gauge data in the study area, and this holds for all temporal scales considered.
(2) The results, eventually, mean that the TMPA TRMM-3B42-V7 product may only be regarded as suitable for further rainfall analyses and subsequent hydrological applications [23,43] in the study region if some improvements, down-scaling, and local calibrations of its output data are carried out first.

Author Contributions: A.J.K. conceived and designed the experiments, conducted the analysis, and is responsible for most of the writing; M.K. helped to develop the idea, supervised the analyses and the writing process, and is responsible for parts of the text; K.M.C. is responsible for parts of the statistical analysis and helped in developing ideas.

Funding: This research received no external funding.

Acknowledgments: Data from the Daily Tropical Rainfall Measurement Mission Project (TRMM) and Others Rainfall Estimate (3B42-V7 derived) used in this study were produced with the Giovanni online data system, developed and maintained by the NASA GES DISC.

Conflicts of Interest: The authors declare no conflict of interest.

Appendix A

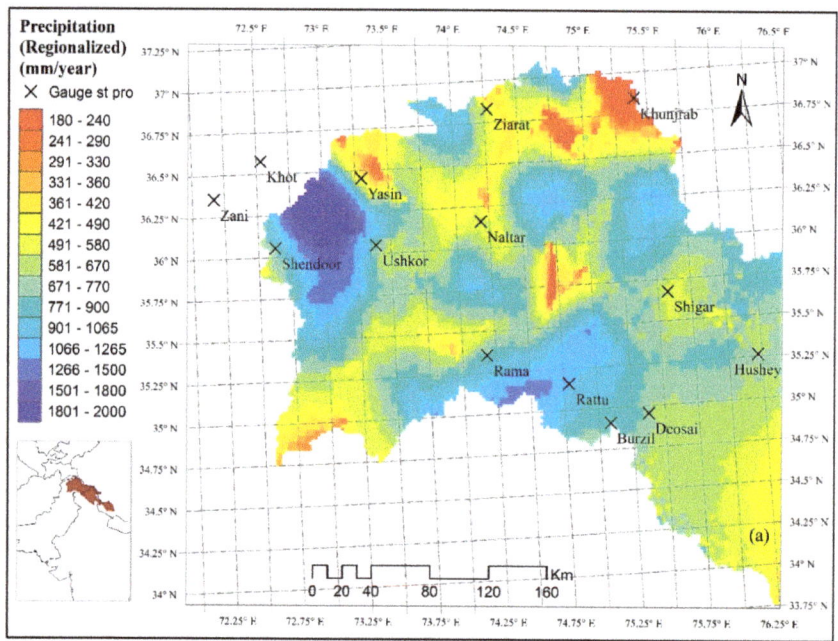

Figure A1. Spatial precipitation regimes in the UIB (adopted from Khan and Koch unpublished).

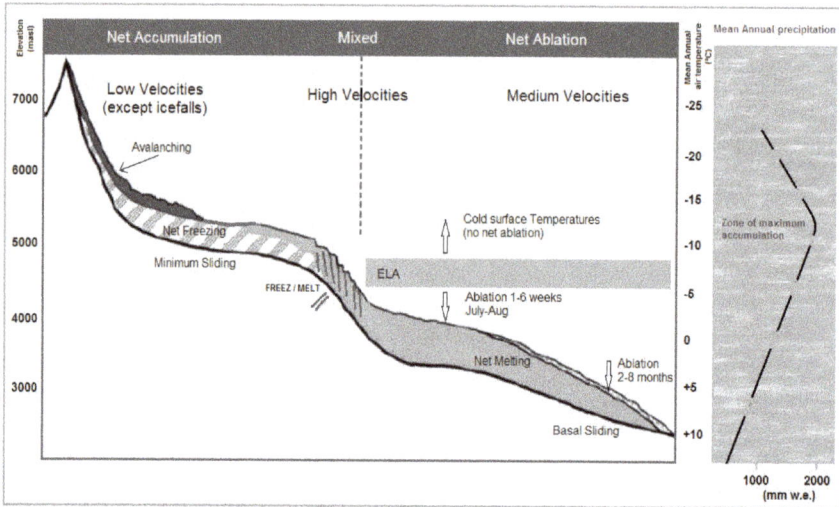

Figure A2. Vertical meteorological and cryspheric regimes in the UIB (modified from Hewitt 2007).

References

1. Scheel, M.L.M.; Rohrer, M.; Huggel, C.; Santos Villar, D.; Silvestre, E.; Huffman, G.J. Evaluation of TRMM Multi-satellite Precipitation Analysis (TMPA) performance in the Central Andes region and its dependency on spatial and temporal resolution. *Hydrol. Earth Syst. Sci.* **2011**, *15*, 2649–2663. [CrossRef]
2. Hasanpour Kashani, M.; Dinpashoh, Y. Evaluation of efficiency of different estimation methods for missing climatological data. *Stoch. Environ. Res. Risk Assess.* **2012**, *26*, 59–71. [CrossRef]
3. Behrangi, A.; Khakbaz, B.; Jaw, T.C.; AghaKouchak, A.; Hsu, K.; Sorooshian, S. Hydrologic evaluation of satellite precipitation products over a mid-size basin. *J. Hydrol.* **2011**, *397*, 225–237. [CrossRef]
4. Pegram, G.; Deyzel, I.; Sinclair, S.; Visser, P.; Terblanche, D.; Green, G. Daily mapping of 24 h rainfall at pixel scale over South Africa using satellite, radar and raingauge data. In Proceedings of the 2nd International Precipitation Working Group (IPWG) Workshop, Monterey, CA, USA, 25–28 October 2004.
5. Ghile, Y.; Schulze, R.; Brown, C. Evaluating the performance of ground-based and remotely sensed near real-time rainfall fields from a hydrological perspective. *Hydrol. Sci. J.* **2010**, *55*, 497–511. [CrossRef]
6. Oke, A.M.C.; Frost, A.J.; Beesley, C.A. The use of TRMM satellite data as a predictor in the spatial interpolation of daily precipitation over Australia. In Proceedings of the 18th World IMACS/MODSIM Congress, Cairns, Australia, 13–17 July 2009.
7. Chiew, F.H.; Vaze, J.; Viney, N.R.; Jordan, P.W.; Perraud, J.M.; Zhang, L.; Teng, J.; Young, W.J.; Peña Arancibia, J.; Morden, R.A.; et al. *Rainfall-Runoff Modelling across the Murray-Darling Basin*; A Report to the Australian Government from the CSIRO Murray-Darling Basin Sustainable Yields Project; CSIRO: Canberra, Australia, 2008.
8. Hughes, D.A. Comparison of satellite rainfall data with observations from gauging station networks. *J. Hydrol.* **2006**, *327*, 399–410. [CrossRef]
9. Joyce, R.J.; Janowiak, J.E.; Arkin, P.A.; Xie, P. CMORPH: A Method that Produces Global Precipitation Estimates from Passive Microwave and Infrared Data at High Spatial and Temporal Resolution. *J. Hydrometeor.* **2004**, *5*, 487–503. [CrossRef]
10. Kubota, T.; Shige, S.; Hashizume, H.; Aonashi, K.; Takahashi, N.; Seto, S.; Hirose, M.; Takayabu, Y.N.; Ushio, T.; Nakagawa, K.; et al. Global Precipitation Map Using Satellite-Borne Microwave Radiometers by the GSMaP Project: Production and Validation. *IEEE Trans. Geosci. Remote Sens.* **2007**, *45*, 2259–2275. [CrossRef]
11. Ushio, T.; Sasashige, K.; Kubota, T.; Shige, S.; Okamoto, K.I.; Aonashi, K.; Inoue, T.; Takahashi, N.; Iguchi, T.; Kachi, M.; et al. A Kalman Filter Approach to the Global Satellite Mapping of Precipitation (GSMaP) from Combined Passive Microwave and Infrared Radiometric Data. *JMSJ* **2009**, *87A*, 137–151. [CrossRef]
12. Aonashi, K.; Awaka, J.; Hirose, M.; Kozu, T.; Kubota, T.; Liu, G.; Shige, S.; Kida, S.; Seto, S.; Takahashi, N.; et al. GSMaP Passive Microwave Precipitation Retrieval Algorithm: Algorithm Description and Validation. *JMSJ* **2009**, *87A*, 119–136. [CrossRef]
13. Turk, F.J.; Rohaly, G.D.; Hawkins, J.; Smith, E.A.; Marzano, F.S.; Mugnai, A.; Levizzani, V. Meteorological applications of precipitation estimation from combined SSM/I, TRMM and infrared geostationary satellite data. In *Microwave Radiometry and Remote Sensing of the Earth's Surface and Atmosphere*; VSP Intl. Sci. Publ.: Zeist, The Netherlands, 2000; pp. 353–363.
14. Huffman, G.J.; Adler, R.F.; Bolvin, D.T.; Nelkin, E.J. The TRMM Multi-Satellite Precipitation Analysis (TMPA). In *Satellite Rainfall Applications for Surface Hydrology*; Gebremichael, M., Hossain, F., Eds.; Springer: Dordrecht, The Netherlands, 2010; pp. 3–22.
15. Huffman, G.J.; Bolvin, D.T.; Nelkin, E.J.; Wolff, D.B.; Adler, R.F.; Gu, G.; Hong, Y.; Bowman, K.P.; Stocker, E.F. The TRMM Multisatellite Precipitation Analysis (TMPA): Quasi-Global, Multiyear, Combined-Sensor Precipitation Estimates at Fine Scales. *J. Hydrometeorl.* **2007**, *8*, 38–55. [CrossRef]
16. Andermann, C.; Bonnet, S.; Gloaguen, R. Evaluation of precipitation data sets along the Himalayan front. *Geochem. Geophys. Geosyst.* **2011**, *12*. [CrossRef]
17. Amir Khan, A.; Pant, N.C.; Ravindra, R.; Alok, A.; Gupta, M.; Gupta, S. A precipitation perspective of the Hydrosphere-cryosphere interaction in the Himalaya. *Geol. Soc. Lond. Spec. Publ.* **2018**, *462*, 73–87. [CrossRef]
18. Hussain, S.; Song, X.; Ren, G.; Hussain, I.; Han, D.; Zaman, M.H. Evaluation of gridded precipitation data in the Hindu Kush–Karakoram–Himalaya mountainous area. *Hydrol. Sci. J.* **2017**, *62*, 2393–2405. [CrossRef]

19. Cheema, M.J.M.; Bastiaanssen, W.G.M. Local calibration of remotely sensed rainfall from the TRMM satellite for different periods and spatial scales in the Indus Basin. *Int. J. Remote Sens.* **2012**, *33*, 2603–2627. [CrossRef]
20. Yatagai, A.; Kamiguchi, K.; Arakawa, O.; Hamada, A.; Yasutomi, N.; Kitoh, A. APHRODITE: Constructing a Long-Term Daily Gridded Precipitation Dataset for Asia Based on a Dense Network of Rain Gauges. *Bull. Am. Meteorl. Soc.* **2012**, *93*, 1401–1415. [CrossRef]
21. Palazzi, E.; Filippi, L.; von Hardenberg, J. Insights into elevation-dependent warming in the Tibetan Plateau-Himalayas from CMIP5 model simulations. *Clim. Dyn.* **2017**, *48*, 3991–4008. [CrossRef]
22. Wijngaard, R.R.; Lutz, A.F.; Nepal, S.; Khanal, S.; Pradhananga, S.; Shrestha, A.B.; Immerzeel, W.W. Future changes in hydro-climatic extremes in the Upper Indus, Ganges, and Brahmaputra River basins. *PLoS ONE* **2017**, *12*, e0190224. [CrossRef] [PubMed]
23. Khan, A.J.; Koch, M. Correction and informed regionalization of precipitation data in a high mountainous region (Upper Indus Basin) and its effect on SWAT-modelled discharge. Unpublished work. 2018.
24. Wilheit, T.T. Some Comments on Passive Microwave Measurement of Rain. *Bull. Am. Meteorl. Soc.* **1986**, *67*, 1226–1232. [CrossRef]
25. Janowiak, J.E.; Joyce, R.J.; Yarosh, Y. A Real-Time Global Half-Hourly Pixel-Resolution Infrared Dataset and Its Applications. *Bull. Am. Meteorl. Soc.* **2001**, *82*, 205–217. [CrossRef]
26. Ali, K.F.; de Boer, D.H. Spatial patterns and variation of suspended sediment yield in the upper Indus River basin, northern Pakistan. *J. Hydrol.* **2007**, *334*, 368–387. [CrossRef]
27. Hewitt, K. Hazards of melting as an option: Upper Indus Glaciers, I&II. *DAWN*, 20 May 2001.
28. Tahir, A.A.; Chevallier, P.; Arnaud, Y.; Neppel, L.; Ahmad, B. Modeling snowmelt-runoff under climate scenarios in the Hunza River basin, Karakoram Range, Northern Pakistan. *J. Hydrol.* **2011**, *409*, 104–117. [CrossRef]
29. Immerzeel, W.W.; van Beek, L.P.H.; Bierkens, M.F.P. Climate Change Will Affect the Asian Water Towers. *Science* **2010**, *328*, 1382–1385. [CrossRef] [PubMed]
30. Wake, C.P. Glaciochemical Investigations as a Tool for Determining the Spatial and Seasonal Variation of Snow Accumulation in the Central Karakoram, Northern Pakistan. *Ann. Glaciol.* **1989**, *13*, 279–284. [CrossRef]
31. Hewitt, K. Glacier Change, Concentration, and Elevation Effects in the Karakoram Himalaya, Upper Indus Basin. *Mt. Res. Dev.* **2011**, *31*, 188–200. [CrossRef]
32. Ali, S.; Li, D.; Congbin, F.; Khan, F. Twenty first century climatic and hydrological changes over Upper Indus Basin of Himalayan region of Pakistan. *Environ. Res. Lett.* **2015**, *10*, 14007. [CrossRef]
33. Hasson, S. Future Water Availability from Hindukush-Karakoram-Himalaya upper Indus Basin under Conflicting Climate Change Scenarios. *Climate* **2016**, *4*, 40. [CrossRef]
34. Singh, P.; Kumar, N. Effect of orography on precipitation in the western Himalayan region. *J. Hydrol.* **1997**, *199*, 183–206. [CrossRef]
35. Dhar, O.N.; Rakhecha, P.R. The effect of elevation on monsoon rainfall distribution in the central Himalayas. In *Monsoon Dynamics*; Lighthill, M.J., Pearce, R.P., Eds.; Cambridge University Press: New York, NY, USA, 1981; pp. 253–260.
36. Dahri, Z.H.; Ludwig, F.; Moors, E.; Ahmad, B.; Khan, A.; Kabat, P. An appraisal of precipitation distribution in the high-altitude catchments of the Indus basin. *Sci. Total Environ.* **2016**, *548–549*, 289–306. [CrossRef] [PubMed]
37. Pang, H.; Hou, S.; Kaspari, S.; Mayewski, P.A. Influence of regional precipitation patterns on stable isotopes in ice cores from the central Himalayas. *Cryosphere* **2014**, *8*, 289–301. [CrossRef]
38. Archer, D. Contrasting hydrological regimes in the upper Indus basin. *J. Hydrol.* **2003**, *274*, 198–210. [CrossRef]
39. Mayor, Y.G.; Tereshchenko, I.; Fonseca-Hernández, M.; Pantoja, D.A.; Montes, J.M. Evaluation of Error in IMERG Precipitation Estimates under Different Topographic Conditions and Temporal Scales over Mexico. *Remote Sens.* **2017**, *9*, 503. [CrossRef]
40. Wilks, D.S. *Statistical Methods in the Atmospheric Sciences. An Introduction*; Academic Press: San Diego, CA, USA, 1995.
41. Wilks, D.S. *Statistical Methods in the Atmospheric Sciences*, 3rd ed.; Elsevier/Academic Press: Amsterdam, The Netherlands, 2011.

42. Palazzi, E.; von Hardenberg, J.; Provenzale, A. Precipitation in the Hindu-Kush Karakoram Himalaya: Observations and future scenarios. *J. Geophys. Res. Atmos.* **2013**, *118*, 85–100. [CrossRef]
43. Khan, A.J. Estimating the Effects of Climate Change on the Water Resources in the Upper Indus Basin (UIB). Ph.D. Thesis, Universität Kassel, Kassel, Germany, 2018.

© 2018 by the authors. Licensee MDPI, Basel, Switzerland. This article is an open access article distributed under the terms and conditions of the Creative Commons Attribution (CC BY) license (http://creativecommons.org/licenses/by/4.0/).

Case Report

Estimating the Future Function of the Nipsa Reservoir due to Climate Change and Debris Sediment Factors

Fotios Maris [1,*], Apostolos Vasileiou [2], Panagiotis Tsiamantas [2] and Panagiotis Angelidis [1]

[1] Department of Civil Engineering, Democritus University of Thrace, 67100 Xanthi, Greece; pangelid@civil.duth.gr
[2] Deparment of Forestry and Management of the Environment and Natural Resources, Democritus University of Thrace, 68200 Orestiada, Greece; apovassi@gmail.com (A.V.); panos-tsiam@windowslive.com (P.T.)
* Correspondence: fmaris@civil.duth.gr; Tel.: +30-254-107-9888

Received: 19 March 2019; Accepted: 24 May 2019; Published: 28 May 2019

Abstract: The constantly growing human needs for water aiming to supply urban areas or for energy production or irrigation purposes enforces the application of practices leading to its saving. The construction of dams has been continuously increasing in recent years, aiming at the collection and storage of water in the formed reservoirs. The greatest challenge that reservoirs face during their lifetime is the sedimentation caused by debris and by the effects of climate change on water harvesting. The paper presents an investigation on the amount, the position and the height of the debris ending up at the Nipsa reservoir. The assessment of the debris volume produced in the drainage basin was conducted by a geographical information system (GIS) based model, named TopRunDF, also used to predict the sedimentation area and the sediment deposition height in the sedimentation cone. The impact of climate change to the reservoir storage capacity is evaluated with the use of a water balance model triggered by the HadCM2, ECHAM4, CSIRO-MK2, CGCM1, CCSR-98 climate change models. The results predict a significant future decrease in the stored water volume of the reservoir, and therefore several recommendations are proposed for the proper future functioning and operation of the reservoir.

Keywords: debris; water balance; climatic change; dam capacity; simulation of sediment transport

---

1. Introduction

Dams are usually constructed to collect and store water in reservoirs for a variety of applications, such as municipal water supplies, energy production and irrigation demand on water coverage. However, the creation of a reservoir entails various risks [1,2]. Proper management and protection of water and soil resources also constitute a major problem.

The concentration of sediments is a piece of information of great worth for many reasons. The production, transportation and deposition of sediment yield are very complicated processes and they vary extremely both in space and time. There are also fluctuations within and between catchments. The deposition of sediments in reservoirs constitutes their greatest threat [3], as it negatively affects their performance due to storage capacity losses, damage to conduits and valves and degraded water quality [4]. Also, the reduction of volume storage capacity may reduce the ability for flood attenuation of the reservoir.

Currently, there is some evidence that the planet is warming up, mainly as a result of human activities producing greenhouse gases [5–7]. Therefore, the estimation of the impact of climate change in managing water resources is more important than ever. The climate change and its impact on extreme hydrological events and generally on water resources constitute nowadays one of the major challenges. Extreme precipitation events are expected to be significantly increased, mainly in relatively wet regions, while the predictions for dry regions are completely different, i.e. prolongation of dry periods [6,7].

The reduction of a reservoir's active storage capacity due to sediment transport and the reduction of water volume in the reservoir affected by climate change are two particularly significant risks [8]. Therefore, it is considered imperative to use sediment yield and water balance models with integrated climate change scenarios in order to avoid any negative economic, environmental and socio-political impact. Thus, in this paper, silting assessment and water balance models with integrated climate change models were used, aiming to estimate the reduction of the Nipsa's reservoir active storage capacity due to sediment transport and climate change. The characteristics of the debris and sedimentation were assessed with the use of the TopRunDF model, while data derived from the HadCM2, ECHAM4, CSIRO-MK2, CGCM1 and CCSR-98 climate models were used to quantify the impact of climate change. Despite the fact that the proposed methodology is applied to a certain reservoir and watershed, the cleanliness of the methodology supports its implementation to all similar basins.

2. Materials and Methods

2.1. Study Area

The study area (Figure 1) is in the northeast part of Greece, in the prefecture of Evros, northeast of the town of Alexandroupoli, near the village of Nipsa. The Nipsa dam is located 20 km northeast of Alexandroupoli, in the Dipotamos area, Loutros torrent. The dam was constructed in 2006 and the purpose of the water basin is to solve the problem of supplying water to the greater area of the Municipality of Alexandroupolis and of the latter's bordering boroughs for the following 40 years at least. The reservoir area equals to 1.1 km^2 and the total storage volume is 13,500,000 m^3. The drainage basin of the reservoir expands over an area of 100 km^2 and the elevation varies from +400 m to +920 m. In the present study, the watershed is divided in 7 sub-basins in order to have a better investigation. The geology of the catchment (Figure 2b) is largely dominated by sedimentary, volcanosedimentary series and volcanic rocks. The basin is mainly characterized by humid continental climate conditions. The mean annual temperature is 15 °C and the mean annual precipitation equals 667 mm. The runoff is collected by the Loutros torrent and is directed as inflow to the Nipsa reservoir. The reservoir is very important for supplying water to the city of Alexandroupolis and its adjacent region, as the treated water is used to cover the needs of about 80,000 habitants. Despite the great importance of the proper functioning of the reservoir for the area, no special work has been done yet. In a bibliographic review made with the use of google scopus, no work was found for this area, contrary to more general work found for the water compartment of Eastern Macedonia and Thrace according to the European Directive 2000/60. According to the Special Water Secretariat Basin Management Plan, the wider area is in a positive water balance, both in the current state and in the future one. At the same time, the found papers on soil loss pertained to an even greater area. In general, the area belongs to the middle soil erosion category. From the studies on dam construction, we found that the available water in the area will be reduced in the future due to climate change, without affecting the operation of the reservoir. We also found that, the debris flow will have no corresponding effects to the reservoir, due to forest vegetation in the area. The advantages of the methodologies we initially presented concern the quantification of the available water and the solid transport as well. At the same time, the simulation of the mass transport phenomenon is extremely important, as the results of generated sediment significantly vary with the volume of sediments that end up in the reservoir.

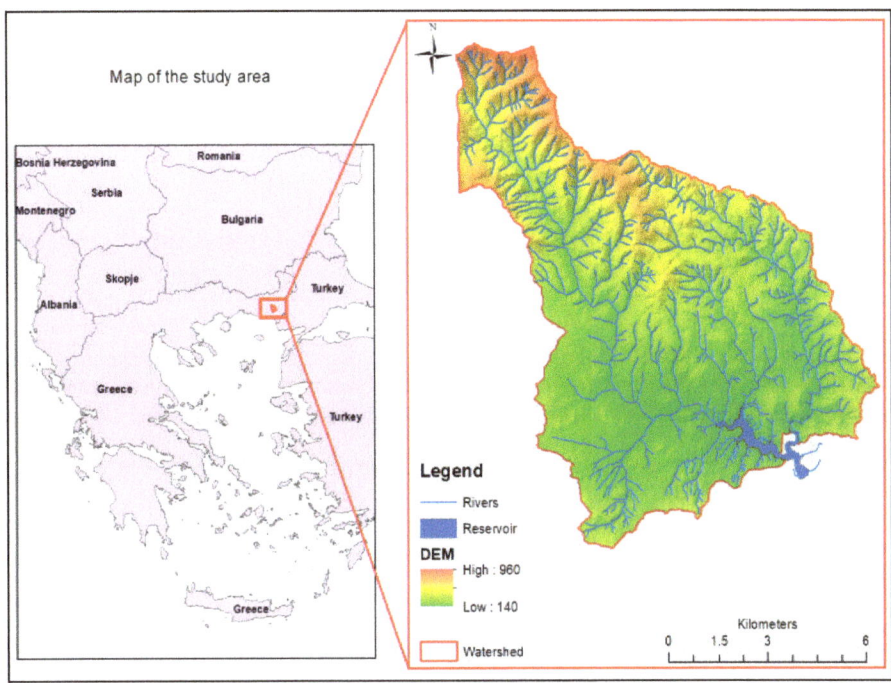

Figure 1. The Nipsa reservoir and the catchment area on the northeast part of Greece.

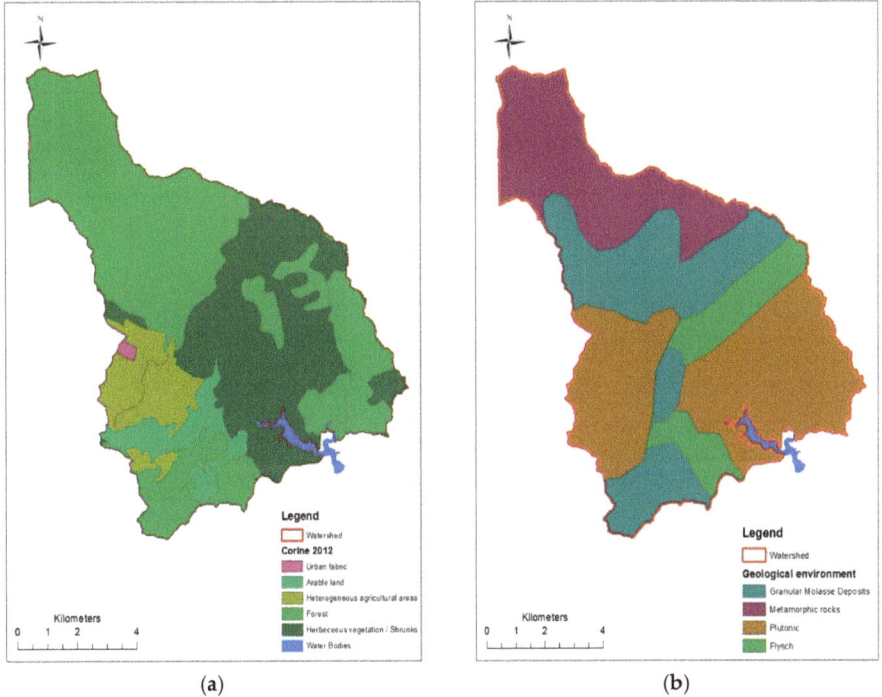

Figure 2. (a) Land use; (b) geological environment.

Methodology

As already mentioned, for the proper operation of the dam it is necessary to consider the available water that will exist in the future in the context of climate change, as well as the tendencies of the reservoir to aggradation. The methodology we followed is divided into two sub-chapters. The first one concerns the tendencies of the reservoir to aggradation. In order to achieve this, the volume of sediment produced in the basin of the area under investigation was calculated using the universal soil loss equation (USLE) and Gavrilovic methods. However, as we are interested in calculating the volume of the deposited materials that end up in the reservoir, we simulated the phenomenon using the TopRunDF model. The next step was to calculate the available water supplying the reservoir in years to come. In order to achieve this, we used the water balance model of the TecnoLogismiki. For the training and validation of the model we used the measured values of the meteorological data, as well as of the water discharge. Then, the timeseries were extended for the next 100 years, and the available water, which is also the baseline scenario, was averaged per month. The extent timeseries were implemented with the various scenarios of climate change and for each scenario, the future available water per month was recalculated anew.

2.2. Sediment Yield Estimation

In the present work, two empirical models were used in order to estimate the soil erosion, namely the USLE [9] and the Gavrilovic method [10].

2.2.1. Universal Soil Loss Equation (USLE)

Source erosion may be estimated using the well-known USLE method. The different factors of the equation have been calculated after processing data coming from small basins from the United States. This certainly constitutes a weakness of the method when it is applied to other regions with different topographic and climatic characteristics [9]. Also, USLE does not include sediment transport in hillslopes and streams and does not behave well in large basins [9]. However, only for estimating the watershed soil erosion, USLE gives a considerably satisfactory preliminary approximation. At this point, it should be pointed out that this specific method is very popular and it provides definitive results. One can ostensively refer to the following papers: Statistical check of USLE-M and USLE-MM to predict bare plot soil loss in two Italian environments, (Vincenzo Bagarello et al., 2018) [11] and Soil Erosion Risk Assessment in Europe (Van der Knijff et al., 2000) [12]. What is more, the European Soil Data Centre has selected the revised USLE method in order to calculate soil erosion in Europe. It is expressed with the following equation:

$$A = R * K * LS * P * C \qquad (1)$$

where A represents the potential long-term average annual soil loss, R the rainfall and runoff factor, K the soil erodibility factor, LS the slope length-gradient factor, P the support practice factor and C the crop/vegetation and management factor.

In this paper, the mean annual rainfall R was calculated in each sub-basin by the Kriging method based on precipitation time-series near the study area. Data from the European Soil Data Center [4,13] were used in order to estimate the K soil erodibility factor. Geographical information systems (GIS) and relevant modules such as the geospatial processing program [14] were utilized for the evaluation of the slope length-gradient factor LS. The support practice factor P was roughly estimated based on observations. The calculation of the crop/vegetation and management factor C was based on the land use database of the European Environmental Agency Corine 2012 (Figure 2a) and was achieved by the use of the use of GIS tools. The torrential streams bank erosion, which is empirically estimated at 20% of the surface erosion, is also added to the total basin erosion [15]. Similar approaches, i.e. modeling soil erosion with GIS, have been also successfully implemented at a river basin in Northern Greece [16].

2.2.2. The Gavrilovic Method

The Gavrilovic method, known as the erosion potential method, was also used in this work as an alternative method to estimate soil erosion. The method was developed for the estimation of the sediment quantity and has been used widely in the Balkan countries since the late 1960s [10] for erosion and torrent-related problems. According to this specific method, the soil erosion is expressed with the following equation:

$$W = T * h * \pi (\sqrt{Z^3}) \tag{2}$$

where W is the average annual production of sediments (m³/year), T is the temperature coefficient (-), h is the mean annual precipitation (mm), Z is the erosion coefficient (-), and F is the area of the catchment (Km²). The temperature coefficient T is given by the following formula:

$$T = \sqrt{(t_o/10 + 0.1)} \tag{3}$$

where t_o is the mean annual temperature (°C) of the basin [17]. The mean annual precipitation h has been calculated with the use of GIS with the same aforementioned method for the calculation of the R factor of the USLE method. The erosion coefficient Z has been evaluated as:

$$Z = x * y * (\varphi + \sqrt{J}) \tag{4}$$

where x is the soil erodibility coefficient (-) expressing the geological medium resistance reduction during erosion, y is the soil protection coefficient (-) depended on the petrological and edaphological composition of the watershed, φ is the coefficient of type and extent of erosion (-), and J is the average slope of the surface of the catchment (%) calculated with the use of GIS.

2.3. Debris-Flow Spread and Deposition Simulation

The GIS based TopRunDF model was selected to predict the possible flow paths on the fan. The model integrates empirical equations with topographical characteristics to predict potential sedimentation areas, as well the sediment deposition height in the sedimentation cone.

2.3.1. TopRunDF Model

The TopRunDF 2.0 model [4,18] is a two-dimensional simulation tool used for the spread of load-bearing materials and for the prediction of the quantity of sediment deposition on the sedimentation cone. Based on the topography of the torrential fan, the model combines a simple flow routing algorithm [19] with the area-volume relation. It is developed with the programming language Visual Basic 6.0 and performs as an integrated executable program using objects of the geographic information system ArcGIS. The input data of TopRunDF are the volume of sediments, the mobility coefficient (K_b), a dimensionless parameter reflecting the flow properties throughout the depositional procedure, the deposition's starting point (fan apex) and the digital terrain model of the watershed (ASCII form, cell size 2,5 m × 2,5 m) [20]. The output results forecast the deposition zones and the height of sediments in these zones. The main advantage of this model lies in the fact that it does not require demanding and time-consuming data, while the accuracy of the results remains notable [21].

2.3.2. Mobility Coefficient

The mobility coefficient (K_b) is a dimensionless parameter. Thus, it should be defined by the user. In the case that a past event has been simulated, it is recommended to estimate the observed K_{bobs} using the empirical relationship:

$$K_{bobs} = B_{bobs} * V_{obs}^{-2/3} \tag{5}$$

where B_{obs} is the observed area of deposition and V_{obs} is the observed volume of the sedimentation.

In this work, the TopRunDF 2.0 model was used to predict potential locations of deposition. Hence, the mobility factor was estimated by the average slope of watercourses S_c, and the average slope of sedimentation cone S_f using the following equation:

$$K_b = 5.075 * S_f^{0.1} * S_c^{-1.68} \tag{6}$$

When the model is used for prediction, it is necessary to take into account the factor of uncertainty. For the selected case study area, this factor is estimated to be equal to 2. The said value is given as a pre-selection by the creators of the model, given the fact that the volume of the sediment is not observed, but calculated using either the USLE or the Gavrilovic method. In our case, other values were also applied, but the trial and error method led us to a factor of uncertainty equal to 2, at which the model behaves in the best way. By using the GIS functionalities, the mean slopes of the streams in the area and the slope of the sedimentation cone were calculated.

2.4. Water Balance Modelling

To quantify the effects of the climate change, the Technologismiki Works 2013 software [22] was selected for modeling the water balance. This software simulates the hydrological cycle of a drainage basin on a monthly time basis. The main input parameters include rainfall, temperature and measured discharge rates. For the calculation of the potential evapotranspiration, the Thornthwaite's method has been implemented, mainly due to lack of detail datasets. In the present work precipitation data, temperature time series and observed runoff data were available for the hydrological years of 2005–2011.

2.5. Climate Change Scenarios

For the comparison of climate change impacts to the water balance in our case study, a reference scenario (without climate change) was created. In particular, synthetic rainfall and temperature series were created to expand the existing time series per 100 years using simple stochastic models [23,24] (autocorrelation, moving averages, etc.) such as AR(1), AR(2), MA(1) and ARMA(1,1). Data series from the climate change scenario HadCM2GGA1, ECHAM4GGA1, CSIRO A1a, CGCM1-A and CCSRGGA1 were retrieved by the climatic models HadCM2, ECHAM4, CSIRO-MK2, CGCM1 and CCSR-98. For each model data set, the changes in precipitation in percentages (%) related to the historical values for the twelve months of the hydrological year were imported to the water balance model. Also, changes in temperatures were given in absolute Celsius degrees (negative sign for a decrease in temperature and positive sign for an increase).

3. Results and Discussion

The visualization of the calculation of sediment discharge using both the USLE and Gavrilovic methods was accomplished with the use of GIS tools, and its results are presented in Figure 3. At first glance from this figure it can be assumed that the amount of sediment transport produced in the water basin is significant and can cause serious problems to the reservoir. But is that the real case? For this question to be answered, the simulation of sediment transport performed with TopRunDF should be taken into account. Initially, in order to have a thorough examination of the drainage network, the basin was divided into 7 sub-basins (Figure 4). In addition, for the limitation of TopRunDF software to be applied per contributor stream, the aforementioned delineation of the watershed is necessary.

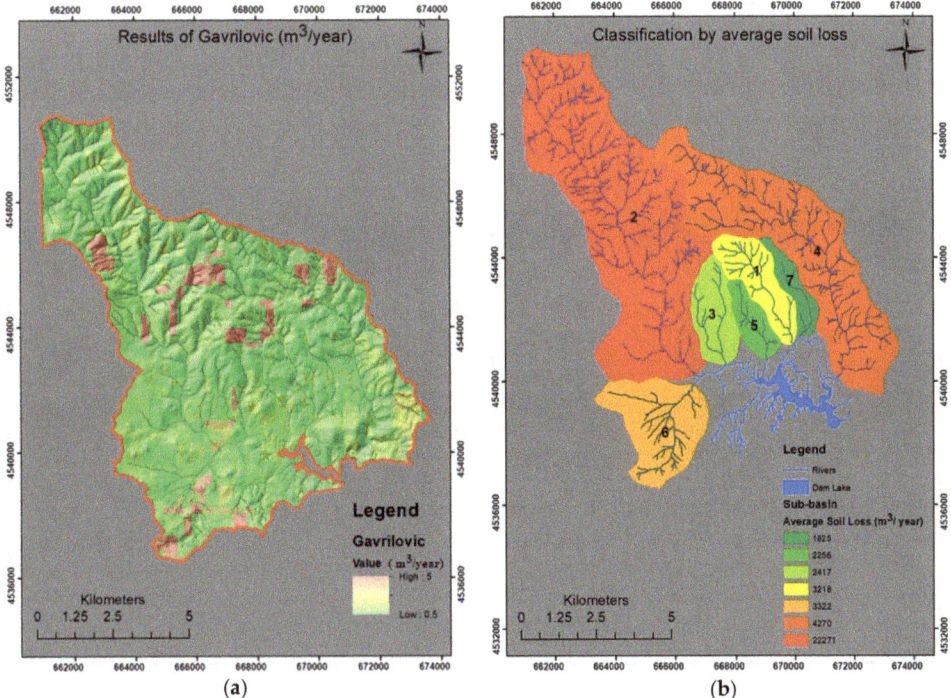

Figure 3. Calculation of sediment discharge by: (**a**) Gavrilovic method and (**b**) classification by average soil.

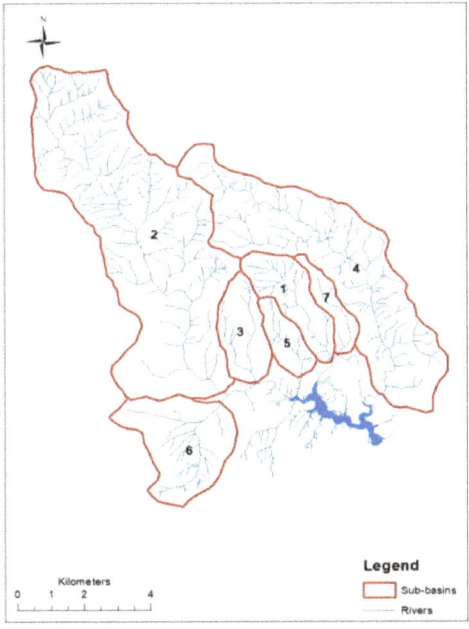

Figure 4. Sub-basins of study area.

The Table 1 below shows the results of the USLE and Gavrilovic method factors as mean values per basin.

Table 1. Values of factors of the Gavrilovic and universal soil loss equation (USLE) methods.

ID	GAVRILOVIC						USLE			
	x	y	φ	J	T	h	R	LS	K	C
1	0,1111	0,62	0,25	13,19	1,505	481,2	381,696	1,334	0,553	0,048
2	0,1516	0,83	0,45	16,39	1,505	481,2	381,696	2,753	0,806	0,026
3	0,2366	0,65	0,35	12,62	1,505	481,2	381,696	1,159	0,596	0,097
4	0,2287	0,71	0,40	14,61	1,505	481,2	381,696	2,172	0,663	0,021
5	0,2913	0,54	0,15	10,48	1,505	481,2	381,696	0,804	0,454	0,130
6	0,2677	0,74	0,20	8,33	1,505	481,2	381,696	0,688	0,695	0,049
7	0,1416	0,67	0,15	11,20	1,505	481,2	381,696	0,887	0,649	0,022

More analytically, in the Gavrilovic method, x is the soil erodibility coefficient expressing the geological medium resistance reduction during erosion, y is the soil protection coefficient which is dependent on the petrological and edaphological composition of the watershed, φ is the coefficient of type and extent of erosion, T is the temperature coefficient (-), and h is the mean annual precipitation (mm). In the USLE method, R is the rainfall and runoff factor, K is the soil erodibility factor, LS is the slope length-gradient factor, and C is the crop/vegetation and management factor. At this point, the importance of some factors, and in particular R, h, T, which are directly related to climate change (temperature and precipitation), should be stressed. Initially, the R factor in the USLE method equals $R = 0.83 * N - 17.7$ (where N is the annual rainfall), while in the Gavrilovic method the corresponding coefficient is the h, which is the annual precipitation and the temperature coefficient T is given by the following formula: $T = \sqrt{(t_0/10 + 0.1)}$ where t_0 is the mean annual temperature (°C). According to climate change models, the annual rainfall is reduced significantly (from 10% up to 40% approximately, depending on the climate scenario), and therefore the volume of the sediment materials will also be decreased. When it comes to the coefficient C, there is the question of whether there will be a change of vegetation due to climate change and how it will affect the sediment discharge. In the study area, due to its geographical location and the tree species, it is expected that climate change will cause small changes, mainly in the expansion of species without any kind of alteration in their protective role of retaining sediment materials [25]. Therefore, we consider that the factor values will remain stable, as they are not dependent on the tree species but on the land use. For a higher accuracy and precision in our conclusions, the future research plans are to run vegetation change models such as maxent [26] and perform statistical analysis of the regional meteorological data in relation to climate change, in order to determine accurately the sediment transport in the future by reducing the bias between observed and climate data.

As is evident in the latter Table 2, the most significant basins referring to debris flow are the basins no. 2 (22,271 m³/year) and no. 4 (4,270 m³/year) (values refer to the average of the two methods and were calculated by using the zonal statistic), while the maximum water discharge values were 87.02 m³/sec and 79.38 m³/sec, respectively.

Due to the importance of these sub-basins, they were selected to be applied to the TopRunDF model. The mobility coefficients were calculated with the use of GIS and were equal to 57 and 49 respectively. The repetition number of the Monte Carlo simulation was defined to 5000 for both of them. For the sub-basin no. 2, the TopRunDF estimated 38,882 possible deposition areas out of which 38,700 were simulated by the model, and the maximum deposition height was 0.57 m. For the sub-basin no.4, the corresponding results were 15,600 possible deposition areas, out of which 15,000 were simulated by the model, and the maximum deposition height was 0.28 m. The spatial representation of the results is depicted in Figure 5, where Figure 5a illustrates the inundated area combined with the overflow possibility of each related cell, while Figure 5b demonstrates the deposition area and the deposition height of each cell.

Table 2. Debris flow values of Gavrilovic and USLE methods and maximum possible discharge.

ID	USLE (m³/Year)	Gavrilovic (m³/Year)	Average (m³/Year)	Maximum Discharge (m³/sec)
1	3150	3286	3218	57.35
2	22959	21583	22271	87.02
3	2129	2705	2417	67.35
4	4150	4390	4270	79.38
5	2350	2162	2256	62.79
6	3522	3122	3322	59.82
7	1928	1722	1825	45.78

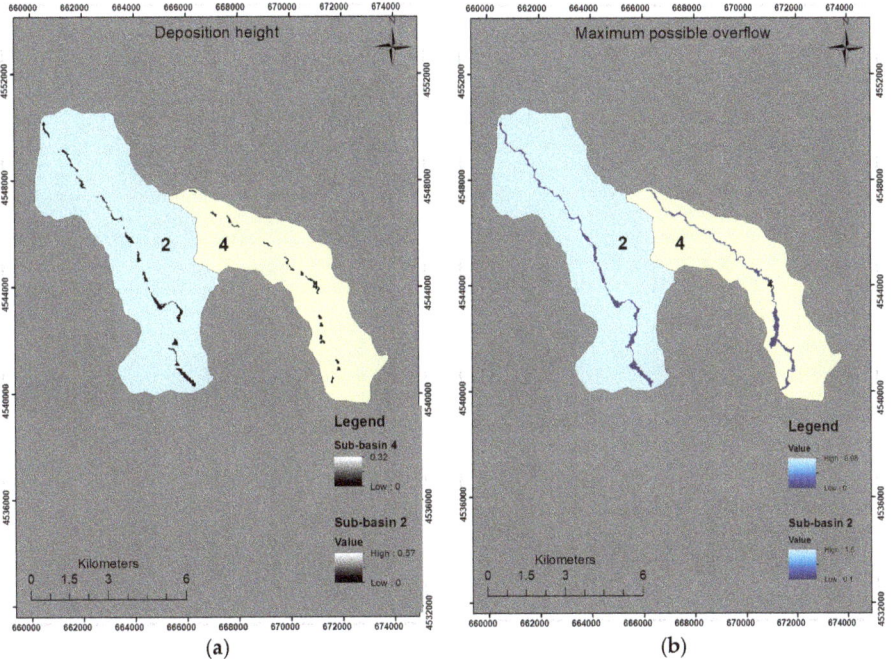

Figure 5. (a) Maximum possible overflow (left), (b) amount of sediment deposition (right).

At this point it should be mentioned that a fire that struck the region in 2011 burned a very small area (1.45 km²) of basin 4, which corresponds to a percentage of 6.45% of the total area. That resulted to an increase of the sediment transfer of 4803 m³/year (about 11%), a fact demonstrating the protective role of vegetation. The new data were introduced to the TopRunDF program and we noticed that the maximum deposition height increased to 0.32 m from 0.28 m, while there were no substantial differences in spatial representation.

Regarding the impact of climate change, the results of the water balance model were taken into account. The first step in the training of the model was the introduction of monthly time series.

For the best calibration of the model, the self-regulation of auto correlation was selected, within limits set by the authors. The limits were: the maximum soil moisture (50–200 mm), the percent of surplus water that flows directly K1 (0.1–2) the percent of groundwater from the previous month K2 (0.1–2), the temperature limit that below percentage is minimum T0 (−2–0), the temperature limit above precipitation that is rain T1 (2–4), the minimum percentage of rain to total precipitation A (0.2–2), the daily factor of snow melting DF (mm deg/day) (0.1–2) and the monthly runoff coefficients (common boundary 0.1–0.6). Based on the results, the mean Nash coefficient was calculated to 0.76 with a maximum of 0.89 and a minimum of 0.7. In addition, the water balance simulation was accomplished with the use of an artificial time series of 100 years, with the input data to be produced by the stochastic models AR (1), AR (2), MA (1) and ARMA (1,1). The use of the Portmanteau test and the inter-comparison of stochastic models led ultimately to the selection of the AR (2) model. Table 3 shows the percentage of increase/decrease of available water that inflows into the reservoir in relation to the reference scenario and the climate change models. Table 4 presents the average monthly volumes of water entering the reservoir, as well as the corresponding monthly volumes for the five climate change scenarios derived from the five climate models.

Table 3. Percentage (%) of increase/decrease of available water for 100 years (with red font is the increase).

Oct	Nov	Dec	Jan	Feb	Mar	Apr	May	Jun	Jul	Aug	Sep	Average	Months/Scenarios
49.0	34.4	55.1	8.7	10.1	62.9	64.1	64.1	63.7	55.4	36.3	26.6	42.75	HadCM2_GGd
28.0	14.2	19.7	2.1	44.9	33.4	26.3	18.4	23.4	29.1	39.9	28.1	25.63	ECHAM 4
26.0	9.7	33.5	1.2	31.4	19.0	18.8	12.3	18.7	26.3	35.3	26.7	21.58	CSIRO-MK2
28.2	12.6	9.0	0.7	29.9	15.1	24.9	2.4	21.1	37.6	61.8	34.2	15.94	CGCM1
18.8	3.7	8.8	3.1	4.7	12.6	15.1	16.4	17.1	14.8	16.0	14.3	12.11	CCSR-98

Table 4. Average water volume in million cubic meters for the base scenario and for the total executed climate change scenarios entering into the reservoir per month.

Oct	Nov	Dec	Jan	Feb	Mar	Apr	May	Jun	Jul	Aug	Sep	Sum	Months/Scenarios
1.40	1.79	2.99	2.55	0.32	1.01	1.22	1.39	0.88	0.68	0.27	0.67	15.17	Base Scenarios
0.71	1.18	1.34	2.55	0.29	0.12	0.12	0.50	0.32	0.32	0.30	0.17	7.92	HadCM2_GGd
1.01	1.54	2.40	2.77	0.18	0.22	0.24	0.26	0.67	0.67	0.48	0.16	10.60	ECHAM 4
1.03	1.62	1.99	0.32	0.22	0.26	0.26	0.28	0.71	0.71	0.50	0.17	8.09	CSIRO-MK2
1.00	1.57	2.72	0.32	0.23	1.16	1.52	1.43	0.69	0.69	0.42	0.10	11.86	CGCM1
1.14	1.73	2.73	2.56	0.31	0.28	0.28	0.27	0.73	0.73	0.58	0.23	11.55	CCSR-98

The annual average water volume entering into the reservoir for the base scenario and for the climate change scenarios is illustrated in Figure 6. The red line represents the volume capacity of the reservoir. According to all executed climate change scenarios, there is a significant reduction in the annual input of water volume in the reservoir. The loss for filling the reservoir on an annual basis ranges from 1.64 m³ × 10^6 (= 13.50 − 11.86) for CGCMI,1, which is the positive scenario of climate change, to 5.58 m³ × 10^6 (= 13.50 - 7.92) for HadCM2_GGd, which is the negative scenario.

In order to investigate the accuracy of the produced results of sediment transport in the basin, a comparison between the simulation outputs and data from the European Soil Data Centre [27] took place and similarities were presented between the two data sets. The simulation results for the sediment transport lead to the conclusion that limited sediment volumes end up in the reservoir. The results produced were also confirmed by the company that manages the dam, as in 10 years of operation the aggradation has reached only 15 cm. This is mainly due to the soil morphology and to the protective role of vegetation in the study area. Therefore, it is clearly proven that the reservoir does not face any danger from soil erosion phenomena in the years to come.

To sum up, the results show that the water availability (the input of water volume into the reservoir) will be significantly decreased in coming years. In the worst case scenario, the reduction will be 42.75%, while in the best case scenario the decrease of water supply will be 12.11%.

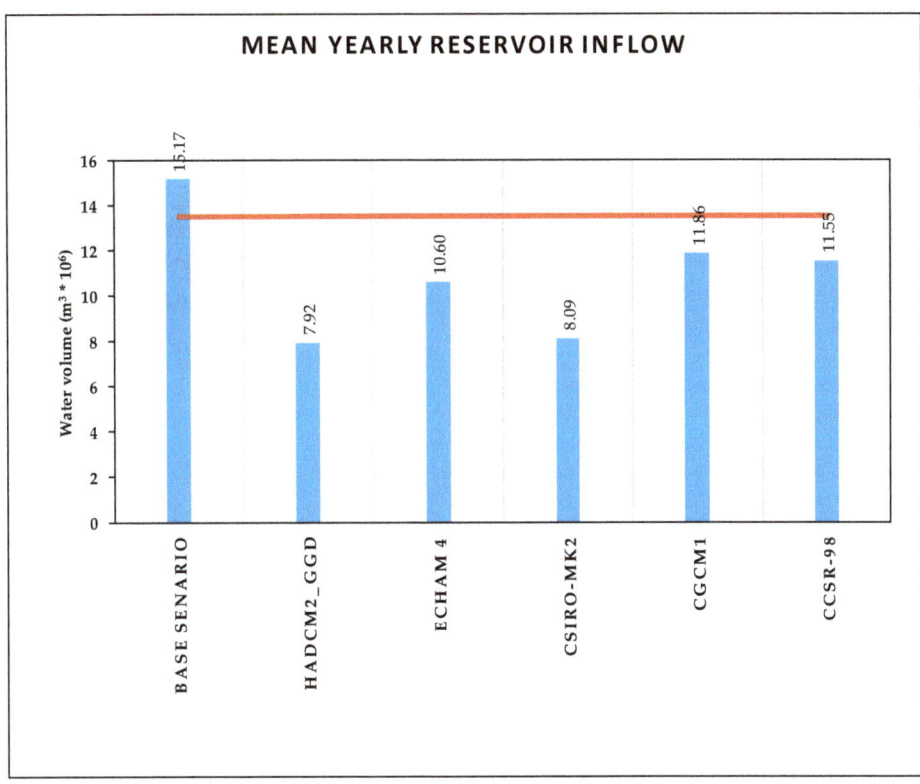

Figure 6. Annual average of input water volume into the reservoir for the base scenario and for the future based on climate change scenarios.

At this point it should be stressed that the values refer to the average yearly values for all years that we simulated the water balance. The trend of the available water is shown in the Table 5 below.

Table 5. Percentage (%) of decrease of available water for 100 years per every 20 years.

2000–2020	2021–2040	2041–2060	2061–2080	2081–2100	Average	Years/Scenarios
39.22	40.89	42.45	44.16	47.03	42.75	HadCM2_GGd
22.79	24.12	25.22	26.77	29.25	25.63	ECHAM 4
20.19	20.57	21.16	22.09	23.89	21.58	CSIRO-MK2
15.37	15.71	15.91	16.08	16.63	15.94	CGCM1
11.5	11.94	12.05	12.2	12.86	12.11	CCSR-98

At the same time, the results of the sediments that end up in the dam, as they emerge from the TopRunDF simulation are very small (about 5 cm/year) and do not affect the dam capacity.

In general, the upcoming climate changes (temperature rise and precipitation reduction) will adversely affect the potential of evapotranspiration and thus, the available soil water. The latter are two essential factors in the determination of the production level of sediments, the stability and the survival of natural ecosystems. Moreover, the expected climate changes will lead to an increase of fire danger [28], reduction of the vegetation density, exposure of soil to surficial and gully erosion, reduction of the basin precipitation buffering capacity and thus, to a reduced production of usable water. In addition, the population growth of the city of Alexandroupolis in the following years, as well as the increase of tourism in the summer months, should lead to a series of measures and actions.

4. Conclusions

At this point, it should be stressed out that the capacity of a dam and its consequent life duration depends on three key factors; the sediment that endures in the reservoir, the available water entering it and its management. In our case, because of the fact that the reservoir is watering the wider area of Alexandroupolis, the management of the dam is rational. As a result, we examined the other factors. Initially, computed the sediment produced in the basin which was calculated using both the USLE and the Gavrilovic methods. Comparing the results with corresponding basins according to the following papers, Assessment of soil erosion intensity in Kolubara District, Serbia [29], Application of USLE, GIS, and Remote-Sensing in the Assessment of Soil Erosion Rates in Southeastern Serbia [30], and according to the review of the Gavrilovic method (erosion potential method) application [31] we also found out the reliability of the methods and the fact that the study area is characterized by medium production of sediment. Using the aforementioned methodologies for soil erosion, we would consider that almost all the bulk of sediment ends up in the reservoir. Therefore, it is necessary to simulate the proposed methodology regarding the control of the tendency of aggradation of the reservoirs, as calculating the deposited materials without the simulation of the phenomenon is not sufficient. The simulation results prove that the reservoir does not face any problem with the deposited materials that enter it. Furthermore, as it has already been mentioned, a fire that struck the region in 2011 burned a very small area (1.45 km^2) of the basin 4, which corresponds to a percentage of 6.45% of the total area. That resulted in an increase in the sediment transfer of 4803 m^3/year (about 11%), a fact demonstrating the protective role of vegetation. The new data were introduced to the TopRunDF software and we noticed that the maximum deposition height increased to 0.32 from 0.28, while in spatial representation there were no substantial differences. Therefore, all necessary measures for the protection of vegetation, such as afforestation and reforestation, sustainable forest management and enforcement of land use, should be taken.

Since both the management of the reservoir and the analysis of the adhesion tendencies lead to the conclusion that the operation of the reservoir is not affected, we focused, on the issue of water availability. The results from the model of water balance depends on each scenario, so that there will be either a great or a small reduction (HadCM2_GGd-CCSR-98). Unfortunately, we cannot know which scenario will prevail. At the same time, we should treat each scenario of climate change as a tool to show us the future tendency, due to its scale and the downscale limitations for its application in the study area, and take all necessary measures and plans to ensure the future operation of the reservoir accordingly as ready management scenarios and preliminary studies to increase its capacity from the neighboring torrents.

Author Contributions: F.M. conceived and designed the research and carried out the analysis of the results. A.V. and P.T. selected the necessary data, performed the numerical simulation and prepared graphs. P.A. analyzed the numerical results and wrote the paper.

Funding: This research received no external funding.

Acknowledgments: We thank the reviewers for their comments and suggestions.

Conflicts of Interest: The authors declare no conflict of interest.

References

1. Sanmartin, J.F.; García, L.A.; Torres, A.M.; Boueno, I.E. Review article: Climate change impacts on dam safety. *Nat. Hazards Earth Syst. Sci.* **2018**, *18*, 2471–2488. [CrossRef]
2. Skoulikaris, C.; Ganoulis, J. Multipurpose hydropower projects economic assessment under climate change conditions. *Fresenious Environ. Bull.* **2017**, *26*, 5599–5607.
3. Kaffas, K.; Hrissanthou, V. Computation of hourly sediment discharges and annual sediment yields by means of two soil erosion models in a mountainous basin. *Int. J. Riv. Basin Manag.* **2017**, *17*, 63–77. [CrossRef]
4. DeNoyelles, F.; Kastens, J.H. Reservoir sedimentation challenges Kansas. *Trans. Kans. Acad. Sci.* **2016**, *119*, 69–81. [CrossRef]
5. Field, C.B.; Barros, V.R.; Dokken, D.J.; Mach, K.J.; Mastrandrea, M.D.; Bilir, T.E.; Chatterjee, M.; Ebi, K.L.; Estrada, Y.O.; Genova, R.C. *Intergovernmental Panel on Climate Change (IPCC). Climate Change: Impacts, Adaptation, and Vulnerability, Part A: Global and Sectoral Aspects, Contribution of Working Group II to the Fifth Assessment Report of the Intergovernmental Panel on Climate Change*; Cambridge University Press: Cambridge, UK, 2014.
6. Kundzewicz, Z.W.; Kanae, S.; Seneviratne, S.I.; Handmer, J.; Nicholls, N.; Peduzzi, P.; Mechler, R.; Bouwer, L.M.; Arnell, N.; Mach, K. Flood risk and climate change: Global and regional perspectives. *Hydrol. Sci. J.* **2014**, *59*, 1–28. [CrossRef]
7. Biao, E.I. Assessing the impacts of climate change on river discharge dynamics in Oueme river basin (Benin, West Africa). *Hydrology* **2017**, *4*, 47. [CrossRef]
8. Spartalis, S.; Iliadis, L.; Maris, F. An innovative risk evaluation system estimating its own fuzzy entropy. *Math. Comput. Model.* **2007**, *46*, 260–267. [CrossRef]
9. Karamage, F.; Zhang, C.; Kayiranga, A.; Shao, H.; Fang, X.; Ndayisaba, F.; Nahayo, L.; Mupenzi, C.; Tian, G. USLE-Based assessment of soil erosion by water in the Nyabarongo River catchment, Rwanda. *Int. J. Environ. Res. Public Health* **2016**, *13*, 835. [CrossRef] [PubMed]
10. Kostadinov, S.; Braunović, S.; Dragićević, S.; Zlatić, M.; Dragović, N.; Rakonjac, N. Effects of erosion control works: Case study—Grdelica Gorge, the South Morava River (Serbia). *Water* **2018**, *10*, 1094. [CrossRef]
11. Bagarello, V.; Ferro, V.; Giordano, G.; Mannocchi, F.; Todisco, F.; Vergni, L. Statistical check of USLE-M and USLE-MM to predict bare plot soil loss in two Italian environments. *Land Degrad. Dev.* **2018**, *29*, 2614–2628. [CrossRef]
12. Van der Knijff, J.M.; Jones, R.J.A.; Montanarella, L. *Soil Erosion Risk Assessment in Europe*; EUR 19044 EN; Office for Official Publications of the European Communities: Luxembourg, 2000; p. 3.
13. Papaioannou, G.; Maris, F.; Loukas, A. Estimation of the erosion of the mountainous watershed of River Kosynthos (in Greek). In Proceedings of the Common Conference of EYE-EEDYP with Title Integrated Water Resources Management under Climatic Changes, Volos, Greece, 27–30 May 2009; pp. 453–460.
14. Ali, S.A.; Hagos, H. Estimation of soil erosion using USLE and GIS in Awassa Catchment, Rift valley, Central Ethiopia. *Geoderma Reg.* **2016**, *7*, 159–166. [CrossRef]
15. Kronvang, B.; Andersen, H.E.; Larsen, S.E.; Audet, J. Importance of bank erosion for sediment input, storage and export at the catchment scale. *J. Soil. Sedim.* **2013**, *13*, 230. [CrossRef]
16. Alexandridis, T.K.; Monachou, S.; Skoulikaris, C.; Kalopesa, E.; Zalidis, G.C. Investigation of the temporal relation of remotely sensed coastal water quality with GIS modeled upstream soil erosion. *Hydrol. Process.* **2015**, *29*, 2373–2384. [CrossRef]
17. Auddino, M.; Dominici, R.; Viscomi, A. Evaluation of yield sediment in the Sfalassà Fiumara (southwestern, Calabria) by using Gavrilović method in GIS environment. *Rend. Online Soc. Geol. Ital.* **2015**, *33*, 3–7.
18. Scheidl, C.; Rickenmann, D. Empirical prediction of debris-flow mobility and deposition on fans. *Earth Surf. Process. Landf.* **2010**, *35*, 157–173. [CrossRef]
19. Kappes, M.S.; Malet, J.P.; Remaître, A.; Horton, P.; Jaboyedoff, M.; Bell, R. Assessment of debris-flow susceptibility at medium-scale in the Barcelonnette Basin, France. *Nat. Hazards Earth Syst. Sci.* **2011**, *11*, 627–641. [CrossRef]
20. Scheidl, C.; Chiari, M.; Rickenmann, D. The use of airborne LiDAR data for the analysis of debris flow events in Switzerland. *Nat. Hazards Earth Syst. Sci.* **2008**, *8*, 1113–1127. [CrossRef]

21. Vasileiou, A.; Maris, F.; Varsami, G. Estimation of sedimentation to the torrential sedimentation fan of the Dadia stream with the use of the TopRunDF and the GIS models. *Adv. Res. Aquat. Environ. Environ. Earth Sci.* **2013**, *3*, 207–214. [CrossRef]
22. TechnoLogismiki Works 2013. Water Budget v9.0—Computational Issue. Available online: http://www.technologismiki.com (accessed on 26 September 2013).
23. Arsenault, R.; Poulin, A.; Côté, P.; Brissette, F. Comparison of stochastic optimization algorithms in hydrological model calibration. *J. Hydrol. Eng.* **2014**, *19*, 1374–1384. [CrossRef]
24. Farmer, W.H.; Vogel, R.M. Research Article: On the deterministic and stochastic use of hydrologic models. *Water Resour. Res.* **2016**, *52*, 5619–5633. [CrossRef]
25. Thuiller, W.; Lavergne, S.; Roquet, C.; Boulangeat, I.; Lafourcade, B.; Araujo, M.B. Consequences of climate change on the tree of life in Europe. *Nat. Int. J. Sci.* **2011**, *470*, 531–534. [CrossRef] [PubMed]
26. Phillips, S.J.; Dudík, M.; Schapire, R.E. Maxent software for modeling species niches and distributions (Version 3.4.1). Available online: http://biodiversityinformatics.amnh.org/open_source/maxent/ (accessed on 28 February 2019).
27. Panagos, P.; Meusburger, K.; Ballabio, C.; Borrelli, P.; Alewell, C. Soil erodibility in Europe: A high-resolution dataset based on LUCAS. *Sci. Total Environ.* **2014**, *479–480*, 189–200. [CrossRef] [PubMed]
28. Wang, X.; Thompson, D.K.; Marshall, G.A.; Tymstra, C.; Carr, R.; Flannigan, M.K. Increasing frequency of extreme fire weather in Canada with climate change. *Clim. Ch.* **2015**, *130*, 573–586. [CrossRef]
29. Belanovic, S.S.; Perović, V.; Vidojević, D.; Kostadinov, S.; Knezevic, M.; Kadović, R.; Košanin, O. Assessment of soil erosion intensity in Kolubara District, Serbia. *Fresenius Environ. Bull.* **2013**, *22*, 1556–1563.
30. Zivotic, L.; Perović, V.; Jaramaz, D.; Djordevic, A.; Petrović, R.; Todorovic, M. Application of USLE, GIS, and remote sensing in the assessment of soil erosion rates in Southeastern Serbia. *Pol. J. Environ. Stud.* **2012**, *21*, 1929–1935.
31. Dragičević, N.; Karleuša, B.; Ožanić, N. A review of the Gavrilovic method (erosion potential method) application. *Gradjevinar* **2016**, *68*, 715–725. [CrossRef]

© 2019 by the authors. Licensee MDPI, Basel, Switzerland. This article is an open access article distributed under the terms and conditions of the Creative Commons Attribution (CC BY) license (http://creativecommons.org/licenses/by/4.0/).

MDPI
St. Alban-Anlage 66
4052 Basel
Switzerland
Tel. +41 61 683 77 34
Fax +41 61 302 89 18
www.mdpi.com

Climate Editorial Office
E-mail: climate@mdpi.com
www.mdpi.com/journal/climate

www.ingramcontent.com/pod-product-compliance
Lightning Source LLC
LaVergne TN
LVHW070603100526
838202LV00012B/546